Acknowledgments

Many people gave generously of their time and expertise in the preparation of this book, and I want to give my most sincere thanks to all of them.

Anne Dow gave helpful comments on several chapters and provided answers to any and all questions about probability, but more importantly, she offered encouragement and support. Eric Hart has for years been a guide to me in teaching and writing about mathematics; for this book, he was always willing to answer even the trickiest question. Joe Tarver patiently read through rough drafts of the many chapters and gave helpful comments and pedagogical advice. Brad Mylett helped me understand various applications of probability, and Giovanni Santostasi assisted with technical points about the normal distribution. Brita Immergut graciously shared her wisdom and experience of teaching and writing about mathematics.

I am deeply grateful to my agent, Elizabeth Frost-Knappman, now retired, whose insight about what makes a good book and what is possible has been invaluable. I thank her for finding an able replacement, Roger S. Williams, who has been of great support in completing this volume.

It has been a great pleasure to work with Emi Smith, Senior Acquisitions Editor at Course Technology, Cengage Learning. She has provided timely assistance with every aspect of the book. Kim Benbow, the project editor for this book, has been a great resource throughout and has given rapid and effective answers to all of my questions. Her keen eye has improved this book in many ways. David Lawrence did a superb job as technical editor, striving for clarity and correctness. I also want to thank Judy Littlefield and Rose Fletcher for their technical assistance.

About the Author

Catherine A. Gorini has taught mathematics at the high school and college levels for over 40 years. She is Professor of Mathematics and Dean of Faculty at Maharishi University of Management. Dr. Gorini received her A.B. in mathematics from Cornell University, M.S. and Ph.D. in mathematics from the University of Virginia, and D.W.P. from Maharishi European Research University. She is the editor of *Geometry at Work* (Mathematical Association of America) and the author of *The Facts On File Geometry Handbook* (Facts On File Science Library, 2009). Her numerous awards for teaching include the Award for Distinguished College or University Teaching of Mathematics from the Mathematical Association of America.

Master Math: Probability

Catherine A. Gorini

Course Technology PTR

A part of Cengage Learning

Australia • Brazil • Japan • Korea • Mexico • Singapore • Spain • United Kingdom • United States

COURSE TECHNOLOGY
CENGAGE Learning

Master Math: Probability

Catherine A. Gorini

Publisher and General Manager, Course Technology PTR:
Stacy L. Hiquet

Associate Director of Marketing:
Sarah Panella

Manager of Editorial Services:
Heather Talbot

Marketing Manager:
Jordan Castellani

Senior Acquisitions Editor:
Emi Smith

Project Editor and Copy Editor:
Kim Benbow

Technical Reviewer:
David Lawrence

Interior Layout Tech:
Judy Littlefield

Cover Designer: Mike Tanamachi

Indexer: Larry Sweazy

Proofreader: Sue Boshers

Library of Congress Control Number: 2010922106

ISBN-13: 978-1-4354-5656-3

ISBN-10: 1-4354-5656-4

Course Technology, a part of Cengage Learning
20 Channel Center Street
Boston, MA 02210
USA

Cengage Learning is a leading provider of customized learning solutions with office locations around the globe, including Singapore, the United Kingdom, Australia, Mexico, Brazil, and Japan. Locate your local office at: **international.cengage.com/region.**

Cengage Learning products are represented in Canada by Nelson Education, Ltd. For your lifelong learning solutions, visit **courseptr.com.**

Visit our corporate website at **cengage.com.**

Printed by RR Donnelley. Crawfordsville, IN. 1st Ptg. 02/2011

Printed in the United States of America
4 5 6 7 8 19 18 17 16 15

To Maharishi Mahesh Yogi

Table of Contents

Introduction

Probability deals with random behavior using mathematics—a task that seems impossible. But, even so, probability finds ways to measure randomness, find order in seemingly chaotic events, and predict the long-term behavior of random processes.

Probability has many uses in today's world. Statistics is the most important application, but areas ranging from communications to genetics to quality control also use probability. Because of its many applications, probability is becoming more and more a part of the core school curriculum.

Master Math: Probability starts at the beginning and explains probability in simple terms. You will find coverage of the background material in set theory, combinatorics, and computational techniques that you will need to understand probability.

This book includes all topics that are usually covered in an introductory course in probability or in a combined probability-statistics course. There are exercises with complete solutions to go with every topic. These are an essential part of the book; how much you get from this book depends completely on working through the exercises on your own or with a fellow student.

You will need a good background in algebra to study probability. You will need to know about functions and their graphs for later chapters. There is a lot of computation in probability, but it is an area where everyone expects you to use a calculator.

I hope that this book will show you the power and beauty of mathematics in dealing with even random outcomes.

Chapter 1

Overview

Probability is the mathematical study of random events; it finds patterns and order in random behavior. Probability is an important mathematical theory because it has many practical applications of its own and is fundamental to the study of statistics.

This chapter gives you an overview of the whole book. First, there is a section giving suggestions for using this book. This is followed by summaries of the content of each chapter.

The mathematical prerequisites for this book are usually covered in high-school mathematics. Computational skill, knowledge of the real number system, algebra, and mathematical reasoning skills are essential. Knowledge of functions (including exponential functions), coordinate geometry, and Euclidean geometry are needed for later chapters. The topics of set theory and combinatorics essential for probability are covered in Chapters 3 and 4. No calculus is required for this book, although some of the theorems in later chapters depend on calculus for their justification. A scientific calculator is necessary and a computer algebra system, such as *Maple*, *Mathematica*, or *Sage*, will be very helpful.

How to Use This Book

This book can be used for self-study to learn probability, as a textbook or a supplement for another textbook, or to review for an examination in probability or a course in statistics.

Each chapter covers one theme, introduced with examples showing computations and applications. You should work through each example by yourself, comparing your work to the text.

At the end of each chapter is a chapter summary, which lists the most important ideas covered in the chapter. Once you have completed the chapter, go over the summary to make sure you have understood each of the ideas. The chapter summary is a good review before starting to work on the exercises. If you are using the book as review, you can look at the summary first; and if you are familiar with some of the topics, you can skip those parts of the chapter.

At the end of every chapter, there are review exercises; these exercises are the most important part of the chapter. Solving each of these exercises on your own is the best way to master the material covered in the chapter. If you find that you cannot solve a problem, go back and review the chapter and make sure you understand the examples. You should look at the answer and solution for an exercise only after you have made a serious effort to solve it yourself. Most mathematics problems can be solved in more than one way, so your method of solution may not match the one given.

See Figure 1.1 for the organization of the chapters in this book. Arrows point from a lower-numbered chapter to a higher-numbered chapter that is dependent on it.

Chapter 2: Introduction to Probability and Its Applications

This chapter gives an informal description of probability that will serve as the intuitive background for further chapters.

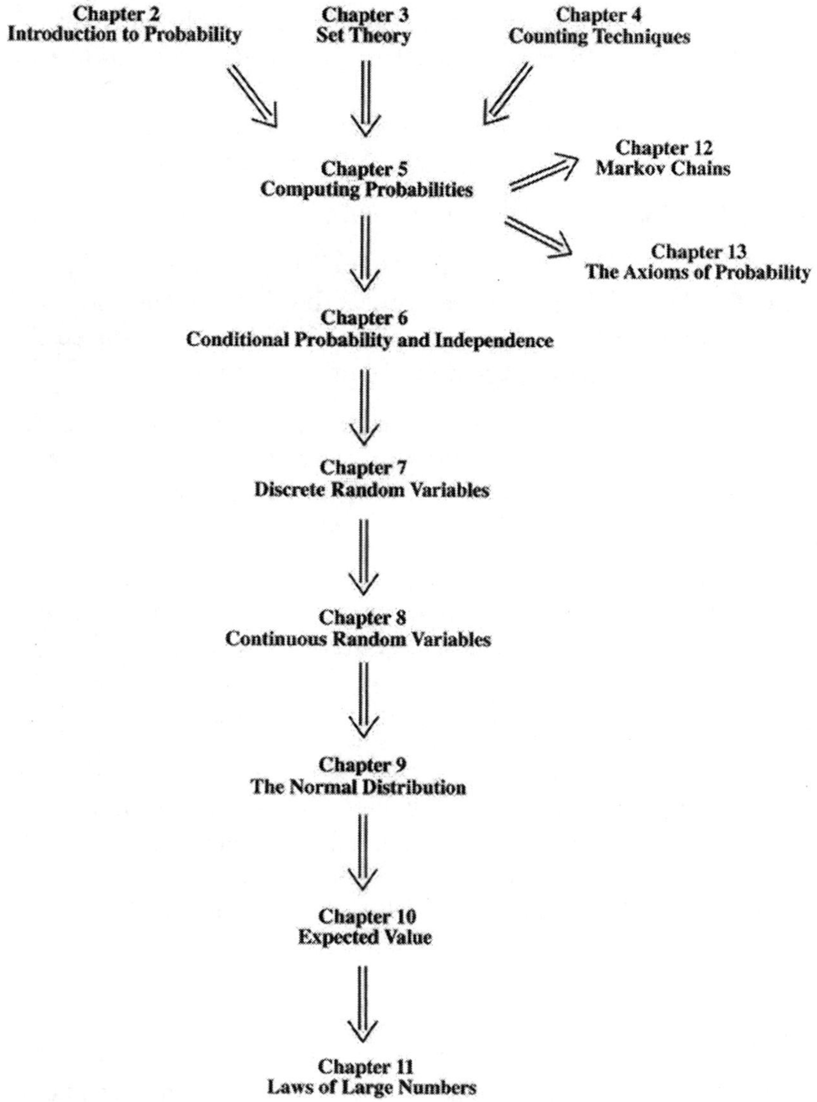

Chapter 2
Introduction to Probability

Chapter 3
Set Theory

Chapter 4
Counting Techniques

Chapter 5
Computing Probabilities

Chapter 12
Markov Chains

Chapter 13
The Axioms of Probability

Chapter 6
Conditional Probability and Independence

Chapter 7
Discrete Random Variables

Chapter 8
Continuous Random Variables

Chapter 9
The Normal Distribution

Chapter 10
Expected Value

Chapter 11
Laws of Large Numbers

Figure 1.1
Chapter dependencies.

We use probability when we perform an experiment or make an observation that has a chance or random outcome. For example, you throw two dice to see what your next move is in a game. Someone deals cards for a hand of bridge after shuffling the deck several times. A random number generator determines the winning number in the Powerball lottery.

These are all random events. This means that they occur by chance, with unpredictable outcomes.

An atom of uranium shoots off a beta particle. It rains. A high-school senior chooses one college rather than another. A light bulb produced by an assembly line is defective. A new drug cures a woman.

Are these random events? In each of these cases, the outcomes could be determined by specific causes, but they may be difficult or impossible to determine. It turns out that these types of events seem, in the long run, to behave as if they were random.

Probability uses mathematics to study these kinds of random events. In other words, probability finds patterns and orderliness in events that appear to occur unpredictably or by chance. While this seems paradoxical, you will see that the tools and procedures covered in this book are effective in finding orderly patterns in events that seem to be random.

Probability is used to study random events, like throwing dice or dealing cards, and gives us an understanding about the long-term behavior of such events. Probability is also useful for studying phenomena, like radioactive decay and the weather, whose behavior appears to be random but may be governed by laws that are unknown or complicated.

Chapter 3: Set Theory

Set theory provides a logical foundation for all of modern mathematics. The basic ideas of set theory are very simple: set, element, and membership. A set is a collection of objects, called elements of the set, and the membership relationship connects an element to a set that it belongs to, or is an element of.

Sets give an orderly way to organize mathematical objects, but their greatest importance is how they can handle infinity. Before the introduction of set theory, mathematicians were restricted when they wanted to study infinite sequences and infinite series and even the real number line because they did not have a systematic way to deal with infinities.

The most important use of set theory in probability is for the sample space, the collection of all possible outcomes of a random experiment, which is a set. An event in probability is just a subset of the sample space, a specific collection of all possible outcomes.

In this chapter, we see the formal definition of sample space and event in terms of sets, elements, and the membership relationship. Sets and their subsets have relationships, such as inclusion, and operations, such as union, intersection, and complementation. These operations give a mathematical structure to set theory like the arithmetic structure of numbers. We also study Venn diagrams, which are pictorial representations of sets that help in the understanding of these operations and relationships.

Chapter 4: Counting Techniques

The most basic definition of probability is in terms of the total number of possible outcomes of a random experiment and the number of these outcomes that are considered to be a "success." To use this definition, we need to do a lot of counting.

Chapter 4 presents the most fundamental counting principles and techniques that are used in probability, including permutations and combinations.

A permutation is an ordering of some or all of the elements of a set. For example, we may want to know how many different ways the president, secretary, and treasurer of a club can be chosen from the members. The order of choosing the members for the offices is important here, because we need to know which member holds which office. On the other hand, a combination is a selection of a number of elements from a set without regard to order. For example, we may want to know how many different ways a nominating committee of three can be chosen from the members of a club.

In this chapter, we also see how Pascal's triangle can be used to determine binomial coefficients.

Chapter 5: Computing Probabilities

This chapter develops many different tools for computing probabilities.

First, we look at rules for numerical computations, including significant digits and rounding. We also look at the rules that help compute more complex probabilities from simpler probabilities: the multiplication rule, the complement rule, and addition rules.

There are three types of problems that are traditionally used over and over as examples in probability—dice, coin tossing, and cards—and Chapter 5 includes many ways to show how to solve such problems. Also in this chapter are several specific examples and applications of probability, including geometric probability, quality control, birthday problems, and transmission errors.

The techniques and examples covered in this chapter will be used over and over in the following chapters.

Chapter 6: Conditional Probability and Independence

This chapter introduces the idea of conditional probability—the probability that an event will occur if we know that some related event has also already occurred. For example, if we throw two dice and we know that doubles have come up, then we know that the sum of the two dice can only be 2, 4, 6, 8, 10, or 12. The probability of any other sum appearing is now equal to 0.

This concept of conditional probability is very important in probability as well as in statistics, where you often have access to some information that is not the information you want or need, but is related to it.

Related to conditional probability is the concept of independent events: two or more events whose outcomes are *unrelated*. For example, the outcomes of the two different dice are independent events, since knowing the number on one die tells us nothing about the number on the other.

In this chapter, you will see the important formulas for conditional probability, including Bayes' formula.

Chapter 7: Discrete Random Variables

In many cases, it is not the outcome of a random experiment itself that we are interested in, but rather something that depends on the outcome. For example, when we roll two dice in a board game, we are interested in the sum of the two dice, not necessarily which numbers appeared.

A random variable is a number that depends on the specific outcome of a given trial. Thus when rolling two dice, the sum of the dice is a random variable associated with the outcome. There are two types of random variables, discrete and continuous. Discrete random variables can have finitely many different values, like the roll of two dice. Continuous random variables take on values in an interval, such as when making a measurement of length or temperature.

A probability distribution is a function that gives probabilities for different values of the random variables. The major distributions covered in this chapter are the joint distribution of two random variables, the binomial distribution, the Poisson distribution, the geometric distribution, the negative binomial distribution, the hypergeometric distribution, and the multinomial distribution.

Because of their importance in probability and its applications, random variables and their associated probability distributions are covered in this and the next two chapters.

Chapter 8: Continuous Random Variables

Continuous random variables take on values in an interval. Measurements on random outcomes, such as the height of a student chosen at random from a class or the lifetime of a light bulb chosen randomly from a production line, are continuous random variables because they take on values in an interval.

Associated with a continuous random variable is a probability density function. The area under the graph of a probability density function is 1, and the function can be used to compute probabilities associated with the random variable.

The continuous random variables considered in Chapter 8 are the uniform distribution, the normal distribution, and the exponential distribution.

Chapter 9: The Normal Distribution

The normal distribution, whose probability density function is a bell-shaped curve, is the subject of this chapter. It is the most important of the continuous distributions because it has so many applications and because it approximates the binomial distribution.

The standard normal distribution is key to understanding all other normal distributions. It has mean 0 and standard deviation 1. Numerical values for the standard normal distribution are available in tables. You can compute probabilities for any other normal distribution from these tables by using a transformation based on the mean and the standard deviation of that distribution.

Chapter 9 reviews the technique of linear interpolation, which finds approximate values for input data not given in a table.

Chapter 10: Expected Value

The expected value, or expectation, of a random variable tells us what average value we can expect the random variable to have in the long run. The expected value is one number that summarizes the distribution of a random variable.

We can compare two random variables by looking at their expected values, and we can judge whether a game of chance is fair or not by looking at the expected value of the game.

To find the expected value for a discrete random variable, compute the product of each value of the random variable times the probability of that value, and then take the sum of these products. The expected value for the discrete random variables belonging to Bernoulli trials, the Poisson distribution, the geometric distribution, the negative binomial distribution, and the hypergeometric distribution are studied.

Formulas for the expected values of continuous distributions, such as the uniform, normal, and exponential distributions are given in this chapter.

Chapter 10 also goes over properties of expected values, including formulas for the expected value of the sum of random variables and the expected value of a multiple of a random variable.

Chapter 11: Laws of Large Numbers

In a random situation, such as flipping a coin, we know that any single outcome is completely random and unpredictable. However, our intuition says that heads and tails should each happen about half the time, *in the long run*. The laws and inequalities discussed in this chapter bridge the gap between knowing that we cannot predict the outcomes of a random experiment and our intuitive feeling that probabilities will tell us what ought to happen.

The most important of these results, the central limit theorem, says that, under certain conditions, random behavior will, in the long run, approximate the normal distribution. This result explains the prevalence of the normal distribution, or the bell-shaped curve, in statistical applications.

All of the inequalities and laws in Chapter 11 are given as probabilities, showing that probability is a tool that can be used to study even itself.

Chapter 12: Markov Chains

A Markov chain is a system that changes from one state or condition to another over time, following probabilistic rules that govern the transitions from each state to the next. One of the best-known examples of a Markov chain is the random walk, where the direction of each step is determined randomly. The random walk can model many real-world situations, including the stock market and the foraging or searching behavior of animals.

Markov chains make extensive use of vectors and matrices, which are reviewed in Chapter 12.

The goal of the study of Markov chains is to predict long-term behavior, which could be a final steady state or a periodic pattern of cycling through different states.

Chapter 13: The Axioms of Probability

Anyone who has studied Euclidean geometry is familiar with the structure first developed by Euclid: postulates or assumptions, undefined terms, definitions, theorems or propositions, and proofs. Since the time of Euclid, mathematicians have regarded this structure to be essential for the validity and secure foundation of mathematics.

Today, set theory is regarded as the foundational theory of mathematics, and set theory itself has an axiomatic presentation.

Chapter 13 describes the axiomatic structure of probability based on the foundation of set theory. The axiomatic structure of probability is necessary for more advanced studies in probability and is interesting in its own right, showing that even the study of random behavior has a firm foundation in logic and set theory and is reliable and valid.

Chapter

Introduction to Probability and Its Applications

The purpose of this chapter is to give an informal introduction to probability. Probability is the mathematical study of random behavior; this means that probability uses the precise and exact methods of mathematics to find patterns in chaotic and unpredictable situations and in outcomes governed by chance.

As early as the sixteenth century, mathematics was used to study games of chance and gambling in order to find patterns that would help gamblers win. By the eighteenth century, probability had become part of mathematics. Finally, in the early twentieth century, when Andrey Kolmogorov gave axioms for probability that were based on set theory and the real numbers, probability was completely integrated into the body of modern mathematics. Today, probability is an important part of mathematics. It has many applications in its own right and as a basis for statistics.

In this chapter, we will see different ways to define probability. The use of mathematical models in the application of probability will be explained. The familiar use of probability in weather forecasting ends the chapter.

2.1 Probability

There are many situations where the outcome of an experiment or procedure seems to be random or due to chance. Flipping a coin, tossing dice, and entering a lottery are common examples that seem to have random outcomes. Other situations, where we do not know all the laws governing the action, such as radioactive decay, errors in communications channels, and genetic inheritance, also appear to be random.

We know that it is impossible to predict the outcome when we flip a coin or toss a pair of dice or examine any other kind of random behavior. However, the theory of probability can help to quantify different random situations, thereby giving a useful understanding of their behavior.

Let's start by looking at a random experiment with many different outcomes, each of which is equally likely to occur. For example, throw two dice. Any one of six numbers can appear on the first die, and any one of six numbers can appear on the second die. If the dice are fair, each of 36 possible outcomes is equally likely.

For this example, suppose that we want doubles and count that as a success. There are 6 of 36 outcomes that are successful. It is reasonable to expect that we will get success (doubles) about 6 / 36, or one-sixth of the time.

This example leads us to a definition of probability. For a random experiment, some of the outcomes are successful and some not. We don't know what will happen after any given trial of the experiment, so we look at the proportion of outcomes that are successful.

If all outcomes of the experiment are equally likely, and some of them are successes, the probability of a successful outcome is defined to be

$$\frac{\text{total number of possible outcomes that are successful}}{\text{total number of possible outcomes}}$$

or simply

$$\frac{\text{number of successes}}{\text{number of outcomes}}$$

For the example of doubles, the probability of a successful outcome is thus

$$\frac{6}{36} = \frac{1}{6}$$

To use this definition of probability, we must have equally likely outcomes and then decide which outcomes are successful. Usually, we count a success to be the outcome that we are most interested in or the outcome that is less frequent. In a board game where doubles gives us an extra turn, it is easy to consider doubles to be a success. Sometimes, the outcome may not really be a "success." For example, an error in data transmission or a defective product from an assembly line might be called a "success," because there are fewer of them and we can more easily count and keep track of them.

We often assume that outcomes are equally likely, even though we could never prove this about real-life situations. So we talk about "perfect coins," "perfect dice," choosing a card from a deck "at random," and the like. Then, we can define the probability that some outcome will happen.

Examples of Probabilities

- When we flip a coin, there are two outcomes, heads or tails. Suppose that we choose heads to be a success. The probability of getting heads when we toss a coin is

$$\frac{1}{2} = 0.5$$

- The probability of getting a 1 when a single die is tossed is 1 / 6, since there are six different outcomes and only one, getting 1, is successful.
- The probability of getting an ace when a card is drawn from a deck of 52 is

$$4 / 52 = 1 / 13 \approx 0.0769$$

since there are 52 different outcomes and only the four aces are successful.

2.2 Ways to Measure Probability

In addition to the fraction representation given in the previous section, there are other common ways to measure the likelihood of an event. Let us take the example of getting a club when we draw a card at random from a deck of 52 cards. The probability is 13 / 52 = 1 / 4 = 0.25 since there are 13 successes out of a total of 52 equally likely outcomes. In this situation, we might say, "The chance of success is one in four."

Probability given as a percent is simply the percent corresponding to the ratio that gives a probability. Thus the probability 0.25 given as a percent is 25%. We say, "There is a 25 percent chance of drawing a club."

Probability is also expressed using odds. The **odds in favor** of an event expresses probability as the ratio of successes to failures. Thus the odds in favor of getting a club when drawing one card from a deck of 52 is

$$\text{number of successes : number of failures} = 1:3$$

Odds are written in various ways, so the odds 1 : 3 can be written as 1–3 or 1 / 3. All of these are read "one to three." We say, "The odds of drawing a club are one to three." **Even odds** means that the odds are 1 : 1.

The **odds against** an event are used less often than odds in favor of an event. The odds against an event gives the ratio of failures to successes. Thus the odds against getting a club when drawing one card from a deck of 52 is

$$\text{number of failures : number of successes} = 3:1$$

Odds are always given as a ratio of integers, and, depending on the context, may not be in lowest terms. So the odds 3 : 1 could be given as 6 : 2 or even 39 : 13.

2.3 Empirical Probability and Subjective Probability

The probabilities in the previous sections are called **theoretical probabilities** because they are based on a theoretical or assumed understanding of the situations. For example, we assume that when we flip a coin, toss dice, or draw a card, all outcomes are equally likely. This is the approach that mathematicians take in the theoretical study of probability, which produces many useful results.

However, if you want to know the probability of getting stuck behind a semi when you arrive at a particular exit ramp near your home, theoretical probability won't work. There is no rule to determine the frequency of semis at the exit ramp. But you can count up how many semis are encountered out of the total number of times you get to the exit ramp. This will give empirical data. Let's consider a semi to be a "success" because that is what you are interested in or focused on, not because you are happy to see one. Count how many times you get to the exit ramp in a particular time period, say a week, as well as how many times you are behind a semi. The probability is given by the ratio

$$\frac{\text{number of semis}}{\text{number of times you arrive at the exit ramp}}$$

So if you arrive at the exit ramp 26 times during a week and are behind semis 5 of those times, the probability of getting behind a semi is $5 / 26 \approx 0.192$.

Such a ratio is called an **empirical probability** because it is based on data collected from repeated trials or experiences. Thus if we make empirical observations and some of them are counted as successes, the probability of success is

$$\frac{\text{number of successes observed}}{\text{number of observations made}}$$

The only way to get an empirical probability is to make observations. More observations will give a more reliable and useful empirical probability.

Empirical probability is used in applications where data can be collected but there is no theoretical basis for a probability. Biologists studying animal behavior, quality control experts inspecting items from a production line, physicists investigating radioactive decay, a baseball batter checking his batting average, and doctors evaluating a new drug all use empirical probabilities.

There is another kind of probability, called **subjective probability**, that refers to a subjective estimation that something will happen.

You might say to a friend, "I have 50–50 odds of getting around to cleaning my room tomorrow." Subjective probability is a good way to express a personal estimation of a situation, and in some cases it may be the only kind of probability that is available. The reliability of a subjective probability depends completely on the prior knowledge and judgment of the individual making the probability.

2.4 How Probability Is Applied

In this section, we will look at how mathematical models are used to apply probability to real-world problems.

Probability describes random behavior, but does anything ever really happen at random? Is there such a thing as luck? These are questions that have been debated for thousands of years. Neither philosophers nor physicists have come to any consensus. Albert Einstein is noted for his insistence that God "does not throw dice" when confronted by the theories of quantum mechanics, which describe nature as functioning in a random way. Someone who has a "chance" encounter that leads to significant changes in their life may be inclined to believe that they just have good luck, or they may want to believe instead that such coincidences are in some way predestined.

While it may be tantalizing to conjecture about such things, mathematicians and scientists who apply probability in a practical way do not need to try to determine whether nature behaves in a random way. This is because many natural systems *seem to behave* as if they are random. Maybe a particular system really is random, or maybe we just don't know in detail the laws governing the system. In either case, probability can be applied effectively.

Flipping a coin seems to have a genuinely random outcome, and we will use flipping coins as an example of a random event over and over. Whenever someone has done an experiment consisting of flipping coins many times, the results conform to probability theory. However, it may be that if you flip a coin and have complete knowledge of the weight of the coin, the position of the coin, the direction and magnitude of the force applied to the coin, the air currents in the room, and so on, you would be able to predict the way the coin would land. So this is an example of a system that *behaves* as if it is random even if we cannot prove that it is random, and we use probability to study flipping coins.

There are many other real-life situations where probability can be applied, even though we might justifiably feel quite certain that there are underlying causes. For example, in the study of Markov chains in Chapter 12, we analyze population changes as if people move from the city to the country *at random*. I think we all realize that a family has very good reasons when they decide to move. But even so, the way the population of a county moves around can be analyzed as if everyone were moving at random.

To understand how to apply probability in a real-life situation, we look at the idea of a mathematical model. Suppose that we have a difficult real-world problem that we want to solve and have collected lots of data. The first step is to convert the real-world problem into a mathematical problem. This mathematical problem is called a **mathematical model** or a **model** because it represents or models the real-life situation.

In creating this mathematical model, we keep only what seems to us to be essential to the real-world problem and ignore details that we think are not relevant to solving the problem. For example, in studying how people move from the city to the country, we are just counting how many people move in a given time period. This is data that we can analyze mathematically. We are not concerned with why they move, whether they will like their new home better, or whether they will miss their friends—this is data that we cannot analyze mathematically.

Once we have reduced the real-world problem to a mathematical problem, we see if we can find a way to solve the mathematical problem. Many of the techniques that you will see in this book can help you solve mathematical problems that model real-world random behavior.

After we have a mathematical solution to the mathematical problem, we interpret the solution in the context of the original real-world problem that we wanted to solve. This process is shown in the diagram in Figure 2.1.

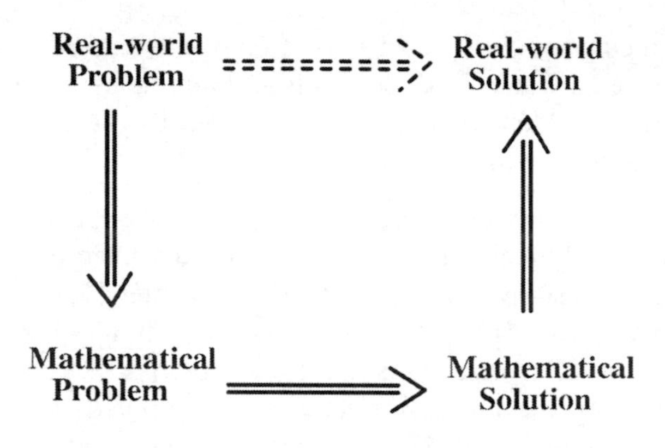

Figure 2.1
How mathematics is applied to real-world problems.

This figure makes you think that using a mathematical model is a roundabout way to solve a problem. However, there are many reasons why using mathematical models is effective.

When a real-world problem is converted to a mathematical problem, we can forget about details that are unimportant or that make the problem more complicated than it needs to be. The real-world problem may be very specific and unique, but the mathematical representation is more general and much simpler. And, of course, it is always a lot easier to solve a simpler problem than a more difficult problem.

The most important reason why using mathematical models is efficient is that, almost always, the mathematical representation of a specific problem is not unique. Once we have converted a real-world problem to a mathematical problem, we can look for similar mathematical problems and then make use of standard tools and general techniques. These mathematical methods may not have been obvious to us when we were looking at the real-world problem.

The mathematical models and techniques that you will be learning in this book are so general, in fact, that they are used in many, many different areas, including

- Medical research
- Quality control
- Risk assessment

- Meteorology
- Statistics
- Stock market analysis
- Quantum mechanics
- Genetics
- Epidemiology
- Computer science
- Communications technology
- Cryptology
- Forensic science
- Thermodynamics
- And even gambling

In the chapters that follow, we will see many real-world situations that can be modeled by random processes. It doesn't really matter to us, as it would to the philosopher, whether or not a given situation is truly random. It only matters whether the random or probabilistic model leads to an effective real-world solution.

And, as it turns out, there are many situations in life that look like they should obey laws, but they behave in such a way that the best models that researchers have found are probabilistic. There are many, many examples of this, including

- Fluctuations in the stock market
- The genes an offspring gets from its parents
- Radioactive decay
- The size of items produced by an assembly line
- Defective items from an assembly line
- Where individuals of a certain species locate their nests
- What the precipitation will be tomorrow
- Errors in a communication channel

The following section will show the familiar use of probability in the daily weather report.

2.5 Weather

To study and predict the weather, meteorologists use many different measurements: temperature, humidity, barometric pressure, wind speed and direction, and so on. They use very sophisticated mathematical models based on calculus. But in the end, they only know that for very similar situations in the past, sometimes it rained and sometimes not. This leads to an empirical determination of how likely precipitation will be in the current situation and a model for weather prediction that uses probability. So when we read a weather report, we find out a chance whether or not there will be precipitation.

Weather reports give the likelihood of rain or snow as a **probability of precipitation** or **PoP**. This number tells the probability that there will be precipitation at any point in the area covered by the weather report. To arrive at the PoP, forecasters look at similar weather conditions in the past and determine an empirical probability of precipitation under those conditions.

The PoP is the product of the probability of precipitation multiplied by the proportion of the area that will experience precipitation if it occurs.

Thus if the forecaster is sure that it will rain, but only in half of the region, the PoP will be 50%. If the forecaster feels that the possibility of rain is 50% and is sure that if it rains at all, every part of the region will get rain, then again the PoP will be 50%. But if the forecaster feels that there is a $5 / 8 = 0.625$, or 62.5%, chance that it will rain in 80% of the region, then the PoP will be $0.625 \times 0.80 = 0.50$, and the report will again give a 50% chance of rain. If the PoP is 100%, it means that the forecaster is sure that it will rain everywhere in the region.

The simplest way to interpret the PoP is that for any point chosen at random in a region, the probability of rain is given by the PoP.

Chapter 2 Summary

Probability is the mathematical study of random behavior. Probability uses the precise and exact methods of mathematics to find patterns in chaotic and unpredictable situations and in outcomes governed by chance.

If all outcomes of an experiment are equally likely, and some of them are successes, the probability of a successful outcome is

$$\frac{\text{number of successful outcomes}}{\text{number of possible outcomes}}$$

This is a **theoretical probability** because it is based on an assumed understanding of the situation.

Probability can be given as the percentage corresponding to the probability ratio. Thus the probability 0.25 given as a percent is 25%.

The **odds in favor** of an event gives the ratio of successes to failures.

Odds are written in various ways, so the odds 1 : 3 can be written as 1–3 or 1 / 3. All of these are read "one to three."

The **odds against** an event gives the ratio of failures to successes.

Odds are always given as a ratio of integers, but they do not have to be in lowest terms. The odds 3 : 1 is the same as 6 : 2 or 39 : 13.

An **empirical probability** is based on data collected from repeated trials or experiences and is

$$\frac{\text{number of successes observed}}{\text{number of observations made}}$$

Subjective probability is a personal estimation of the likelihood of an outcome.

A **mathematical model** is a mathematical problem that represents or models a real-life situation.

Mathematical models are effective because

- They are simpler than the real-world situations they model.
- They allow you to use standard mathematical techniques.
- You can use the same models for many different real-world problems.

Weather reports give a **probability of precipitation** or **PoP**, which tells the probability that there will be precipitation at any random point in the area covered by the weather report.

Chapter 2 Practice Problems

1. You choose one card from a standard deck of 52.

a. What is the probability of getting an ace?

b. Express this probability as odds.

c. Express this probability as a percent.

d. What is the probability of getting a heart?

e. Express this probability as odds.

f. Express this probability as a percent.

g. What is the probability of getting a red card?

h. Express this probability as odds.

i. Express this probability as a percent.

2. You buy a lottery ticket and are told that the odds of winning are 1 : 16. What does that mean?

3. You throw a single die and count 1 or 2 as a success.

a. What is the probability of success as a decimal?

b. What is the probability of success as a percent?

c. What is the probability of success given as odds?

4. Suppose your computer checks your email every five minutes. In a two-hour period, you receive eight notifications that you've got mail. What is the empirical probability that you get mail in a five-minute period?

5. The weather report for your location says "CHANCE OF SNOW SIXTY PERCENT." What does this mean?

6. For the following claims, say whether it is theoretical probability, empirical probability, or subjective probability.

a. The probability of getting two heads in a row when flipping a fair coin

b. The probability of getting doubles with loaded dice

c. The odds of getting a ticket to the concert before all the tickets are sold out

d. The chance of getting a winning lottery ticket

e. The chance of finding my homework assignment in my backpack

f. The probability of finding five eggs in a clutch of robin's eggs

g. The chance of getting a 22 on the next bingo call

h. The chance of a defective cell phone coming off the assembly line

i. The chance of your number coming up on a roulette wheel

Answers to Chapter 2 Practice Problems

1. For each of these outcomes, there are 52 possible successes.

a. $\dfrac{\text{number of successes}}{\text{number of possible outcomes}} = \dfrac{4}{52} = \dfrac{1}{13} \approx 0.0769$

b. 4 : 48 or 1 : 12

c. 7.69%

d. $\dfrac{\text{number of successes}}{\text{number of possible outcomes}} = \dfrac{13}{52} = \dfrac{1}{4} = 0.25$

e. 13 : 39 or 1 : 3

f. 25%

g. $\dfrac{\text{number of successes}}{\text{number of possible outcomes}} = \dfrac{26}{52} = 0.5$

h. 1 : 1

i. 50%

2. It means that, on average, if you purchased many tickets, you would win once for every 16 times you lost.

3. a. $\dfrac{\text{number of successes}}{\text{number of possible outcomes}} = \dfrac{2}{6} \approx 0.333$

b. $0.333 \approx 33\%$

c. 2 : 4 or 1 : 2

4. In a two-hour period there are $2 \times 12 = 24$ five-minute periods. You get mail in eight of them. The empirical probability of getting mail is

$$\frac{\text{number of successes}}{\text{number of trials}} = \frac{8}{24} = \frac{1}{3} \approx 0.333$$

5. This means that the probability of snow at any place chosen at random in the region is 0.60. This number is the product of the probability of snow anywhere in the region times the part of the region where snow is likely to fall, based on data from the past conditions.

6. **a.** Theoretical because it is based on a perfect coin.

   b. Empirical because we don't know how the dice have been loaded, and we have to throw the dice many times to determine the probability of getting doubles.

   c. Subjective because it is based on a personal or intuitive judgment.

   d. Theoretical because it is based on printed odds on the ticket.

   e. Subjective because it is based on personal judgment.

   f. Empirical because it is based on counting many clutches.

   g. Theoretical because it is based on the knowledge of how many numbers have yet to be called and the equal chances for getting each number.

   h. Empirical because it is based on factory data.

   i. Theoretical if the roulette wheel is fair, or empirical if the house has fixed the wheel.

Chapter

Set Theory

Set theory gives a language for talking about collections of objects. It is very important in probability because a sample space is the set of all possible outcomes of an experiment, and events are subsets of the sample space. Events and their relationships are studied using the operations and relationships of set theory.

This chapter will introduce ideas from set theory that are used in the study of probability.

3.1 Sets

A **set** is a collection into a whole of definite, distinct objects. The objects in a set are called the **elements** of the set. Saying that the elements must be definite means that we must be explicit when describing the elements of a set. The condition that the elements are distinct means that the elements must be different—no element can be listed more than once in the set.

The elements of a set are listed between brackets, so $A = \{1, 2, 3\}$ is the set that is named by the letter A and that contains the three elements 1, 2, and 3. To show that a number is an element of a set, we use the symbol $\in$, which means "is an element of." Thus we write $1 \in \{1, 2, 3\}$

or $1 \in A$ to indicate that 1 is an element of the set A. We can also write $\notin$ to mean "is not an element of." So $4 \notin \{1, 2, 3\}$ or $4 \notin A$ says that 4 is *not* an element of the set A.

Examples of Sets

- **{1, 2, 3, 4}** This set has four elements.
- **{1, 2, 3, 4, . . .}** The dots indicate that all integers after 4 are in the set. This set has infinitely many elements.
- **{all positive even integers}** This set has infinitely many elements, 2, 4, 6, and so on.
- **{$x \mid x$ is a positive even integer}** This way of writing a set is called **set builder notation**. The symbol | is read "such that." This is the same set as in the previous example.

3.2 Venn Diagrams

Venn diagrams show relationships between sets in picture form. They are named after the British logician John Venn, who introduced them in 1881. A rectangle is used to represent the universe of all the elements we are looking at, and circles are used to represent sets. The individual elements can be all written in if you like.

If we have elements 1, 2, 3, 4, and 5, then the Venn diagram of set $A = \{1, 2, 3\}$ can be given as one of the two versions shown in Figure 3.1. Sometimes the set under consideration is shaded, as shown in Figure 3.2. The name of the set can be put inside the circle representing the set or else outside and close to the circle.

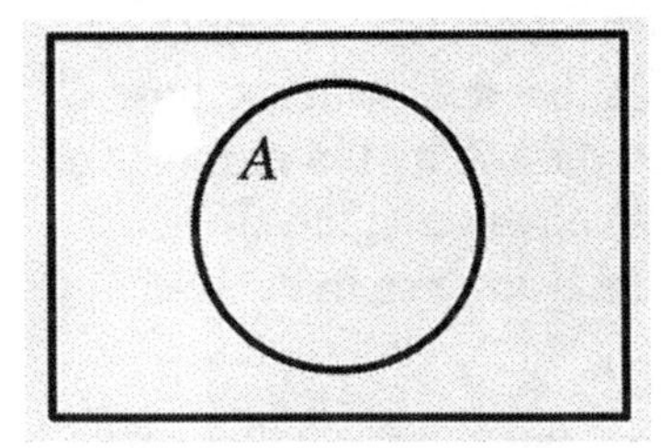

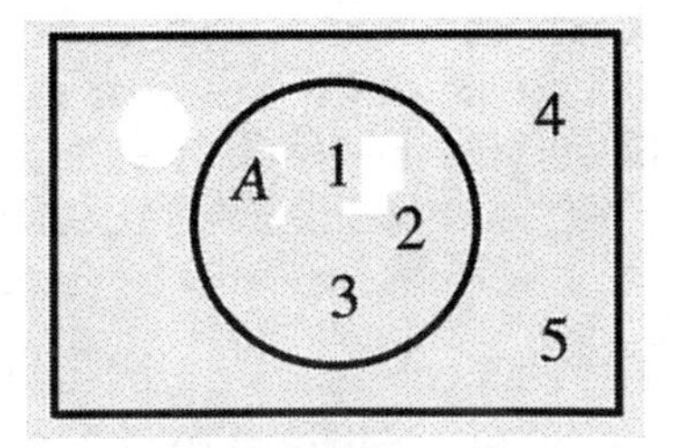

Figure 3.1
Two Venn diagrams showing the set {1, 2, 3} contained in the universe {1, 2, 3, 4, 5}.

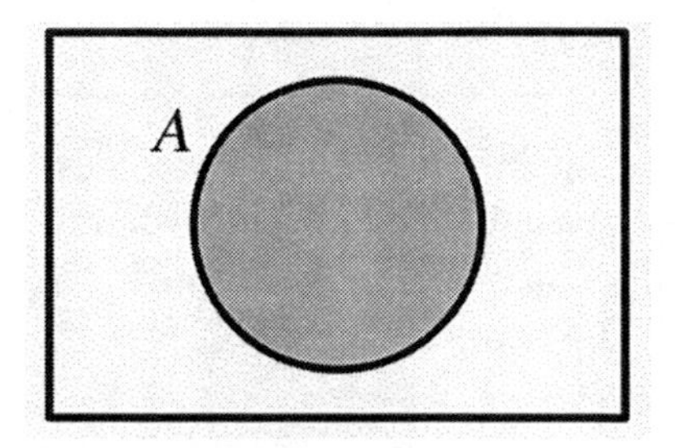

Figure 3.2
Venn diagram with the set *A* shaded.

3.3 Subsets

If all elements of a set A belong to another set B, we say that A is a **subset** of B, and we write $A \subseteq B$. The symbol $\subseteq$ means "is a subset of." We also say that the set A is **included** in the set B. The Venn diagram in Figure 3.3 shows inclusion of the set A in the set B.

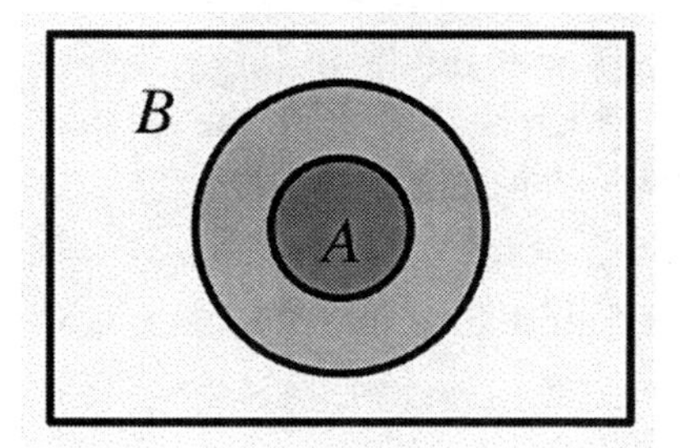

Figure 3.3
A subset *A* of the set *B*.

A set is considered to be a subset of itself, and we write $A \subseteq A$. A subset that is *not* the whole set is called a **proper subset**, and we use the symbol $\subset$ when we want to emphasize this fact. The symbol $\subset$ means "is a proper subset of."

3.4 Sample Spaces and Events

The sample space, as we saw in Chapter 2, contains all possible outcomes for some experiment or observation that we are interested in. The different elements of the sample space are the specific outcomes. In this context, we say that an element of the sample space is a **sample point**.

Examples of a Sample Point

- If we roll a single six-sided die, the sample space is {1, 2, 3, 4, 5, 6} and 6 is one sample point.
- If we choose a day of the week at random, the sample space is {Sunday, Monday, Tuesday, Wednesday, Thursday, Friday, Saturday} and Tuesday is one sample point.
- If we flip two coins, the sample space is {HH, HT, TH, TT} and TH is a sample point.

An **event** is any subset of the sample space. For example, rolling an even number with one die is the event {2, 4, 6}, which is a subset of the sample space in {1, 2, 3, 4, 5, 6} given previously.

If an event consists of only one outcome or element, it is a **simple event** or **elementary event**. If an event consists of more than one outcome, it is a **compound event**.

3.5 Some Special Sets

The set that has no elements is called the **empty set** or the **null set**, and it is denoted by { } or ∅. The set that includes all elements that we are interested in is called the **universal set**, the **universe of discourse**, the **universe,** or **U**. In a Venn diagram, the outer rectangle represents the universal set because it contains all the elements in a given example. In probability, the universal set is the sample space.

3.6 Operations with Sets

Just as numbers have operations, like addition and multiplication, sets also have operations: union, intersection, complementation, difference, and symmetric difference. These operations on sets help us understand how different events in a sample space are related to one another.

The **union** of two sets is the set that contains all of the elements in one or the other of the sets. The symbol for union is ∪. For example, $\{1, 2, 3\} \cup \{2, 3, 4, 5\} = \{1, 2, 3, 4, 5\}$. The union of two sets A and B is shaded in the Venn diagram in Figure 3.4.

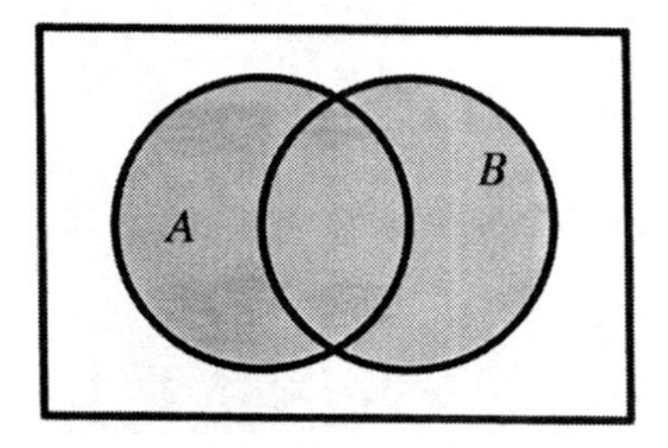

Figure 3.4
The union of sets *A* and *B*.

Examples of a Union

- If one die is thrown, the union of the event that an even number appears with the event that a number larger than three appears is the event $\{2, 4, 5, 6\}$.
- If one card is drawn from a deck of 52, the union of the event that a diamond is drawn with the event that a heart is drawn is the event that a red card is drawn.

The **intersection** of two sets is the set that contains all of the elements that are in both of the sets. The symbol for intersection is ∩. For example, $\{1, 2, 3\} \cap \{2, 3, 4, 5\} = \{2, 3\}$. If the intersection of two sets is the empty set, the two sets are said to be **disjoint**. Events corresponding to disjoint sets are said to be **mutually exclusive**.

In a Venn diagram, the intersection of two sets is the region where the circles representing the sets overlap. The intersection of sets A and B is shaded in the Venn diagram shown in Figure 3.5; two disjoint sets C and D are shown in Figure 3.6.

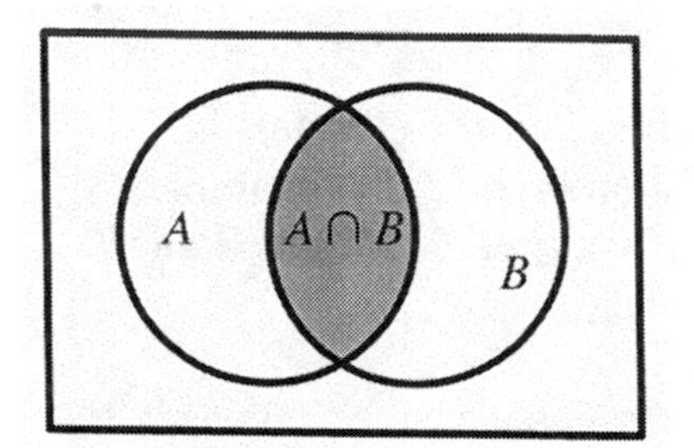

Figure 3.5
The intersection of sets A and B.

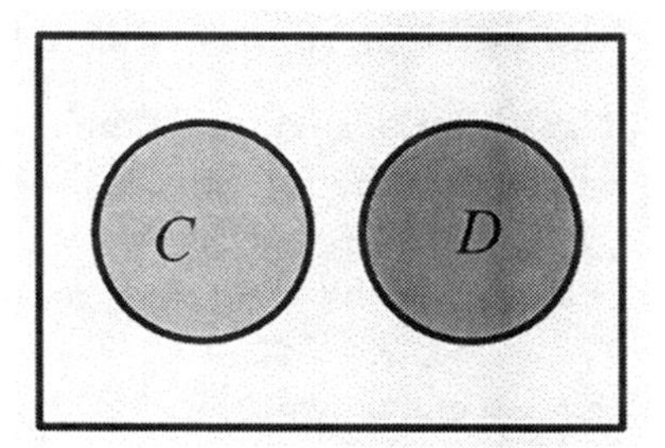

Figure 3.6
Two disjoint sets C and D whose intersection is the null set.

Examples of Intersections

- If one die is thrown, the intersection of the event that an even number appears with the event that a number larger than 3 appears is the event $\{4, 6\}$.
- If one card is drawn from a deck of 52, the intersection of the event that a red card is drawn with the event that an ace is drawn is the event of drawing a red ace, which is the set {A♦, A♥}.

The **complement** of a set A is the set of all the elements in the universal set that do *not* belong to A. The complement of the set A is written A^C. In order to find a complement, you must know the universal set. With respect to the universal set $\{1, 2, 3, 4, 5, 6\}$, the complement of the set $\{1, 2, 3\}$ is the set $\{4, 5, 6\}$. The complement of the empty set $\varnothing$ is the universal set **U** and the complement of the universal set **U** is the null set $\varnothing$.

In the Venn diagram in Figure 3.7, the complement of set A is the region outside of the circle representing set A.

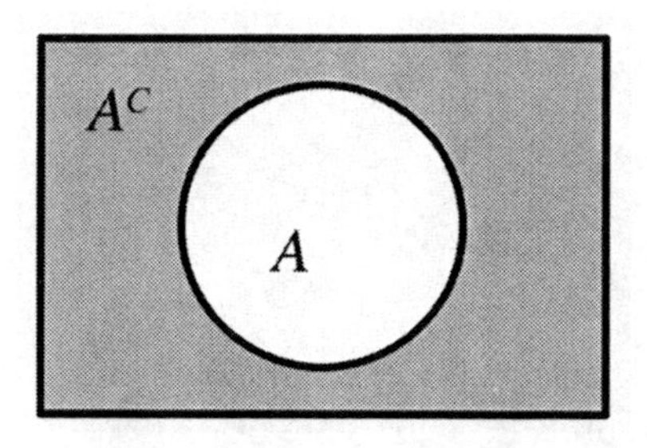

Figure 3.7
The complement A^C of set A.

Examples of Complements

- If one die is thrown, the complement of the event that an even number appears is the event that an odd number appears.
- If one card is drawn from a deck of 52, the complement of the event that a red card is drawn is the event that a black card is drawn.

The **difference** $A \setminus B$ of two sets is the set of all elements that are in set A but not in set B. This means that we are taking away each element of A that is also in B. If A and B are disjoint sets, $A \setminus B = A$. For example, $\{1, 2, 3\} \setminus \{2, 3, 4, 5\} = \{1\}$ and $\{1, 2, 3\} \setminus \{4, 5\} = \{1, 2, 3\}$.

In the Venn diagram for $A \setminus B$, the region inside A that is outside B is shaded, as shown in Figure 3.8.

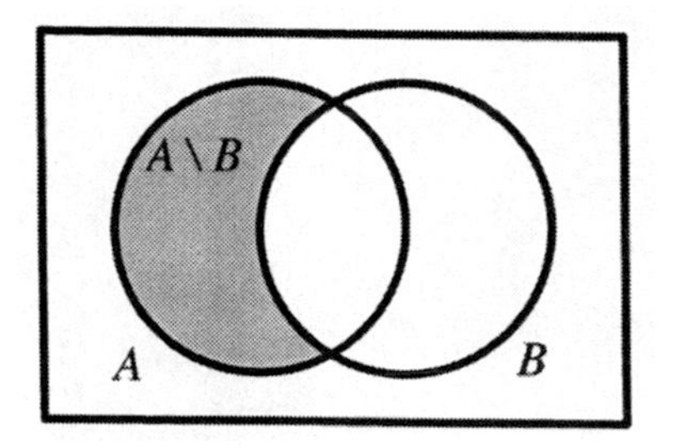

Figure 3.8
The difference $A \setminus B$ of sets A and B.

Examples of Differences

- If one die is thrown, the difference of the event {2, 4, 6} that an even number appears minus the event {4, 5, 6} that a number greater than three appears is the event {2}.
- If one card is drawn from a deck of 52, the difference of an event that an ace appears minus the event that a red card appears is the set of black aces {A♣, A♠}.

The **symmetric difference** $A \oplus B$ of two sets is the set of all elements that are in set A and in set B but not in their intersection. Another way to think of this is that the symmetric difference $A \oplus B$ is the set of elements in A that are not in B along with the elements of B that are not in A. For example, $\{1, 2, 3\} \oplus \{2, 3, 4, 5\} = \{1, 4, 5\}$.

If A and B are disjoint sets, the symmetric difference $A \oplus B$ is the same as the union $A \cup B$.

The Venn diagram of $A \oplus B$ has two regions shaded, the elements of A that are not in the intersection $A \oplus B$ and the elements of B that are not in the intersection $A \oplus B$, shown in Figure 3.9.

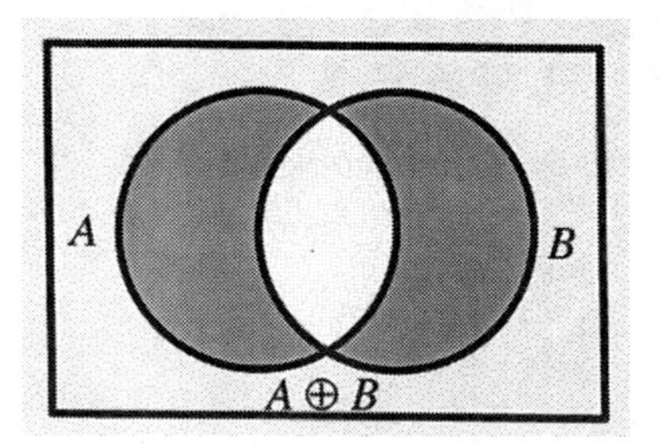

Figure 3.9
The symmetric difference $A \oplus B$ of sets A and B.

Examples of Symmetric Differences

- If one die is thrown, the symmetric difference of the event {2, 4, 6} that an even number appears with the event {4, 5, 6} that a number greater than three appears is the event {2, 5}.
- If one card is drawn from a deck of 52, the symmetric difference of an event that an ace appears with the event that a diamond card appears is the set {A♥, A♣, A♠, K♦, Q♦, J♦, 10♦, 9♦, 8♦, 7♦, 6♦, 5♦, 4♦, 3♦, 2♦}.

3.7 Properties of Set Operations

Set operations have properties like those of the arithmetic operations. Addition and union of sets are both operations of putting things together, and they turn out to have similar properties. It is not obvious why multiplication and intersection of sets should be alike, but it turns out that they also have similar properties. These properties are very useful to simplify expressions and computations.

Examples of properties in arithmetic are commutativity and distributivity. Addition and multiplication are **commutative** because $a + b = b + a$ and $a \times b = b \times a$ for all numbers. Multiplication is **distributive** over addition because $a \times (b + c) = a \times b + a \times c$ for all numbers.

The numbers 0 and 1 are important in arithmetic because they are identity elements—they don't change numbers that they operate on. In addition, we have $0 + a = a$ and in multiplication, $1 \times a = a$. The null set $\varnothing$ plays a role like 0 since $A \cup \varnothing = A$, and the universal set $\mathbf{U}$ or the sample space S plays a role like 1 since $A \cap \mathbf{U} = A$.

Parentheses are used to clarify an expression involving set operations in the same way that they are used to clarify arithmetic or algebraic expressions. The expression $2 + 4 \times 5$ is ambiguous because you don't know whether to perform the addition first or the multiplication first. By putting in parentheses, you can distinguish between $(2 + 4) \times 5 = 30$ and $2 + (4 \times 5) = 22$. In set theory, $A \cup B \cap C$ can be different, depending on whether union or intersection is done first, so parentheses are necessary.

By convention, complementation is done before the other operations. It is not necessary to use parentheses in an expression like $A \cup B^C$ to mean the intersection of the set A with the complement of the set B. However, it is necessary to use parentheses in the expression $(A \cup B)^C$ to mean that the union must be performed before the complementation.

The most important properties of set operations are given below. The events A, B, and C are all taken from the same universe **U**.

- **Commutativity.** This law says that the order of the operation doesn't matter.
  - Commutativity of Union: $A \cup B = B \cup A$
  - Commutativity of Intersection: $A \cap B = B \cap A$
- **Associativity.** This law says that in a sequence of two or more instances of the same operation, it doesn't matter which one is done first. In arithmetic, addition and multiplication are associative since $(a + b) + c$ is the same as $a + (b + c)$ and $(a \times b) \times c$ is the same as $a \times (b \times c)$.
  - Associativity of Union: $(A \cup B) \cup C = A \cup (B \cup C)$
  - Associativity of Intersection: $(A \cap B) \cap C = A \cap (B \cap C)$
- **Distributivity.** In arithmetic, multiplication is distributive over addition, $a \times (b + c) = a \times b + a \times c$ for all numbers, but addition is not distributive over multiplication since, for example, $3 + (2 \times 5)$ is not the same as $(3 + 2) \times (3 + 5)$. Set theory is different, and we have two distributivity laws: intersection is distributive over union, and union is distributive over intersection.
  - $A \cap (B \cup C) = (A \cap B) \cup (A \cap C)$
  - $A \cup (B \cap C) = (A \cup B) \cap (A \cup C)$
- **Identity.** In addition, 0 acts as an identity since $0 + a = a$ and in multiplication, 1 acts as identity since $1 \times a = a$. In set theory, the empty set acts as identity for union, and the sample set S or universal set **U** acts as identity for intersection.
  - $\varnothing \cup A = A$
  - $\mathbf{U} \cap A = A$

- **Idempotence.** A number is an idempotent if the number operated on itself gives the number back again. In addition, 0 is idempotent since $0 + 0 = 0$ and in multiplication 1 is idempotent since $1 \times 1 = 1$. In set theory, every set is idempotent with respect to both union and intersection.
  - $A \cup A = A$
  - $A \cap A = A$
- **Involution.** The operation of complementation in set theory behaves somewhat like finding the additive or multiplicative inverse of a number—the inverse of the inverse of a number is the number itself. For addition, $-(-a) = a$, and for division, $1 / (1 / a) = a$.
  - $(A^C)^C = A$
- **Complements.** These laws show how sets and their complements behave with respect to each other.
  - $A \cup A^C = \mathbf{U}$
  - $A \cap A^C = \varnothing$
  - $\mathbf{U}^C = \varnothing$
  - $\varnothing^C = \mathbf{U}$
- **DeMorgan's Laws.** These laws show how complementation interacts with the operations of union and intersection.
  - $(A \cup B)^C = A^C \cap B^C$
  - $(A \cap B)^C = A^C \cup B^C$

Chapter 3 Summary

Set theory is the language used in probability to talk about sample spaces and events.

A **set** is a collection into a whole of definite, distinct objects.

The objects in a set are called the **elements** of the set.

The symbol $\in$ means "is an element of."

Venn diagrams show the relationships between sets in picture form. A rectangle represents the universe of all the elements under consideration, and circles represent sets.

If all of the elements of a set A belong to another set B, then A is a **subset** of B. The symbol $\subseteq$ means "is a subset of."

A subset that is *not* the whole set is a **proper subset**. The symbol $\subset$ means "is a proper subset of."

A **sample space** is the set of all possible outcomes for some experiment or observation.

An element of a sample space is a **sample point**.

An **event** is any subset of a sample space.

A **simple event** or **elementary event** contains only one element or outcome.

A **compound event** contains more than one element or outcome.

The **empty set** or the **null set** has no elements and is denoted by { } or $\varnothing$.

The set that includes all elements that we are interested in is the **universal set** or the **universe of discourse** and is denoted **U**. In probability, the sample space S is the universal set.

The **union** of two sets is the set that contains all of the elements in one or the other of the sets. The symbol $\cup$ means "union with."

The **intersection** of two sets is the set that contains all of the elements that are in both of the sets. The symbol $\cap$ means "intersected with."

If the intersection of two sets is the empty set, the two sets are said to be **disjoint**.

The **complement** of a set A is the set of all the elements in the universal set that do *not* belong to A. The complement of the set A is written A^C.

The **difference** $A \setminus B$ of two sets is the set of all elements that are in set A but not in set B.

The **symmetric difference** $A \oplus B$ of two sets is the set of all elements that are in set A and in set B but not in their intersection.

Properties of Set Operations

- Commutativity
  - Commutativity of Union: $A \cup B = B \cup A$
  - Commutativity of Intersection: $A \cap B = B \cap A$

- Associativity
  - Associativity of Union: $(A \cup B) \cup C = A \cup (B \cup C)$
  - Associativity of Intersection: $(A \cap B) \cap C = A \cap (B \cap C)$
- Distributivity
  - $A \cap (B \cup C) = (A \cap B) \cup (A \cap C)$
  - $A \cup (B \cap C) = (A \cup B) \cap (A \cup C)$
- Identity
  - $\varnothing \cup A = A$
  - $\mathbf{U} \cap A = A$
- Idempotence
  - $A \cup A = A$
  - $A \cap A = A$
- Involution
  - $(A^C)^C = A$
- Complements
  - $A \cup A^C = \mathbf{U}$
  - $A \cap A^C = \varnothing$
  - $\mathbf{U}^C = \varnothing$
  - $\varnothing^C = \mathbf{U}$
- DeMorgan's Laws
  - $(A \cup B)^C = A^C \cap B^C$
  - $(A \cap B)^C = A^C \cup B^C$

Chapter 3 Practice Problems

1. Suppose three coins are tossed.
   - **a.** Give the sample space as a set S.
   - **b.** Give the event "more heads than tails" as a set E.
   - **c.** Give the event "not all the same" (or "not all heads and not all tails") as a set F.
   - **d.** Give the union of the sets E and F.
   - **e.** Give the intersection of the sets E and F.
   - **f.** Give the complement of the set E.
   - **g.** Describe the complement of the set E using words.

2. Suppose that one card is drawn from a deck of 52. Let A be the event of drawing a face card, B be the event of drawing a heart, C be the event drawing a red card, and D be the event drawing a black card.
   - **a.** How many different events are there?
   - **b.** Give an example of an elementary event.
   - **c.** Give an example of a compound event.
   - **d.** Give as a set the event $B \cup C$.
   - **e.** Give as a set the event $C \cup D$.
   - **f.** Give as a set the event $A \cap B$.
   - **g.** Give as a set the event $B \cap C$.
   - **h.** Give as a set the event $C \cap D$.
   - **i.** Give as a set the event D^C.
3. Draw a Venn diagram of three sets A, B, and C with the property that $A \subset B$ and $B \subset C$.
4. What is the difference of a set A and its complement?
5. Compare the Venn diagrams for $A \setminus B$ and $B \setminus A$. Are they the same or different?
6. **a.** Draw a Venn diagram of $(A \cup B) \cap C$.
   - **b.** Draw a Venn diagram of $A \cup (B \cap C)$.
   - **c.** Are these diagrams the same or different?
   - **d.** Give examples of three sets A, B, and C with the property that $(A \cup B) \cap C$ and $A \cup (B \cap C)$ are different.

Answers to Chapter 3 Practice Problems

1. **a.** S = {HHH, HHT, HTH, HTT, THH, THT, TTH, TTT}
   - **b.** E = {HHH, HHT, HTH, THH}
   - **c.** F = {HHT, HTH, HTT, THH, THT, TTH}
   - **d.** $E \cup F$ = {HHH, HHT, HTH, HTT, THH, THT, TTH}
   - **e.** $E \cap F$ = {HHT, HTH, THH}
   - **f.** E^C = {HTT, THT, TTH, TTT}
   - **g.** "More tails than heads" or "fewer heads than tails" are both correct.

2. **a.** 52
   b. Any set containing one card will be an elementary event; for example, {2♥} is an elementary event.
   c. Any set containing more than one card will be a compound event; for example, {2♥, 10♠, K♦} is a compound event.
   d. $B \cup C = D$, is the set of all red cards.
   e. $C \cup D$ is the set of all 52 cards.
   f. $A \cap B =$ {K♥, Q♥, J♥}.
   g. $B \cap C = B$, the set of all hearts.
   h. $C \cap D = \varnothing$.
   i. $D^C = C$.

3.

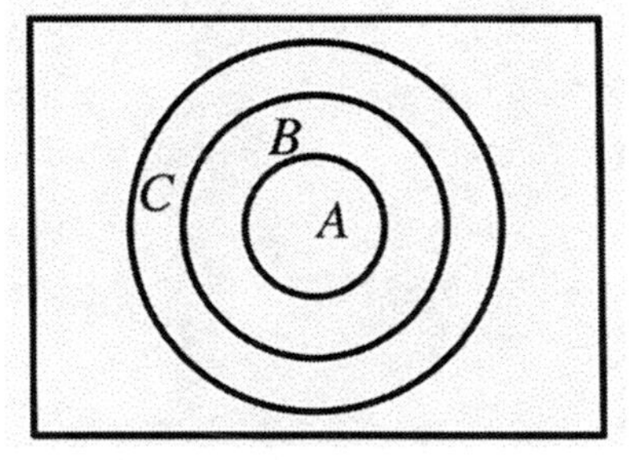

Figure 3.10
The Venn diagram for sets A, B, and C such that $A \subset B$ and $B \subset C$.

4. The set A.
5. See Figure 3.11 for the Venn diagram of $A \setminus B$ and see Figure 3.12 for the Venn diagram of $B \setminus A$. The Venn diagrams for $A \setminus B$ and $B \setminus A$ are different.

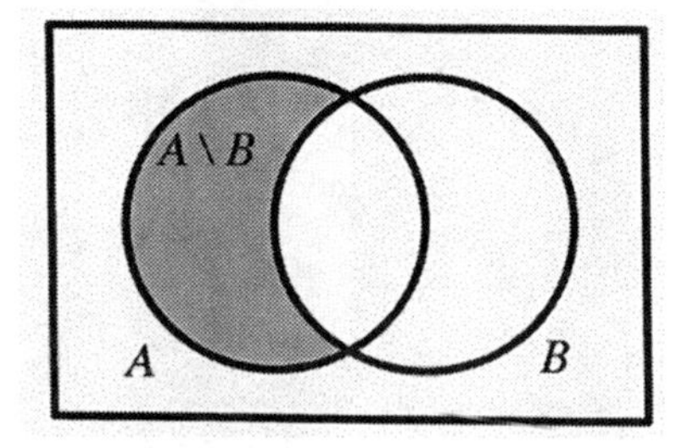

Figure 3.11
The Venn diagram for $A \setminus B$.

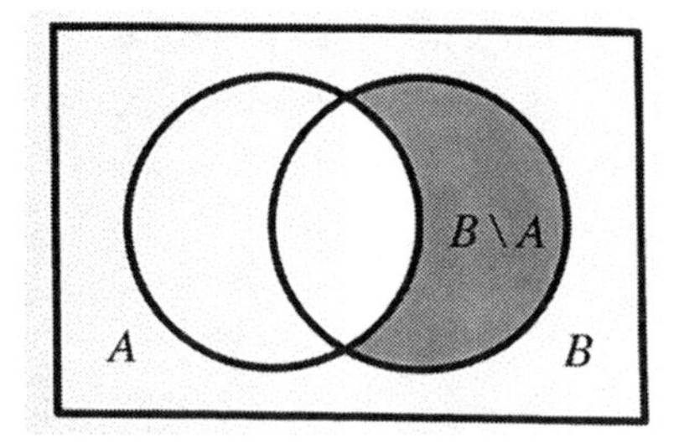

Figure 3.12
The Venn diagram for $B \setminus A$.

6. a.

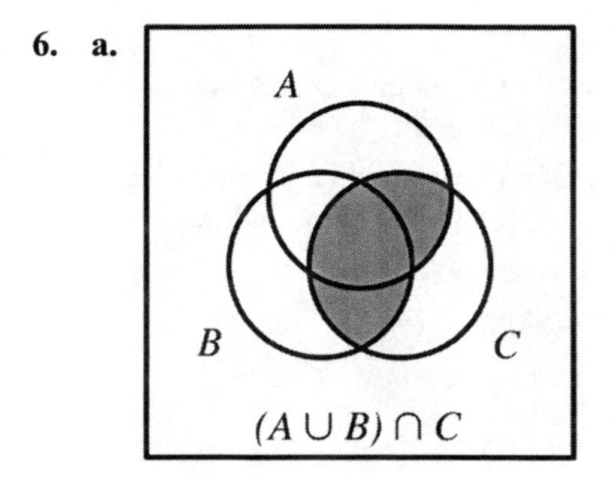

Figure 3.13
$(A \cup B) \cap C$.

b.

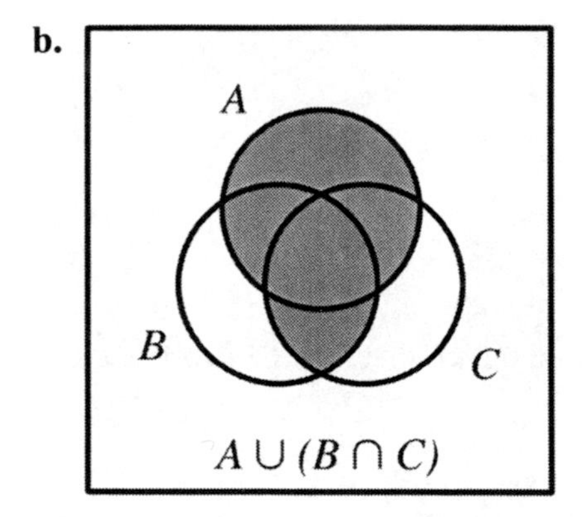

Figure 3.14
$A \cup (B \cap C)$.

c. These are different.

d. There are many examples. One example is $A = \{1, 2, 3\}$, $B = \{2, 3, 4, 5\}$, and $C = \{4, 5, 6, 7\}$.

Chapter 4

Counting Techniques

We saw in Chapter 2 that the probability of an event is the number of equally likely outcomes in the event divided by the number of all possible outcomes. This means that it is very important to be able to count up how many outcomes are possible and how many outcomes belong to the event that you are interested in.

This chapter introduces the basic techniques for counting and shows how to use them.

4.1 Counting Principles

In this section, we will look at the Product Rule and the Sum Rule. The Product Rule is used when making several choices simultaneously or when making a sequence of choices, one after the other. The Sum Rule is used when making one choice from a collection of different sets of possibilities.

Product Rule for Counting

Suppose you have three tops and two pair of pants. How many different outfits do you have? Each top can be worn with each pair of pants, so there are six different outfits. There are two ways to picture this, a grid (or table) and a tree diagram.

The grid or table in Figure 4.1 shows all the different possibilities. There is a column for each different top, and there is a row for each different pair of pants. The pictures in the grid show the different outfits. From this picture, it is easy to see that there are 3×2, or 6, different outfits.

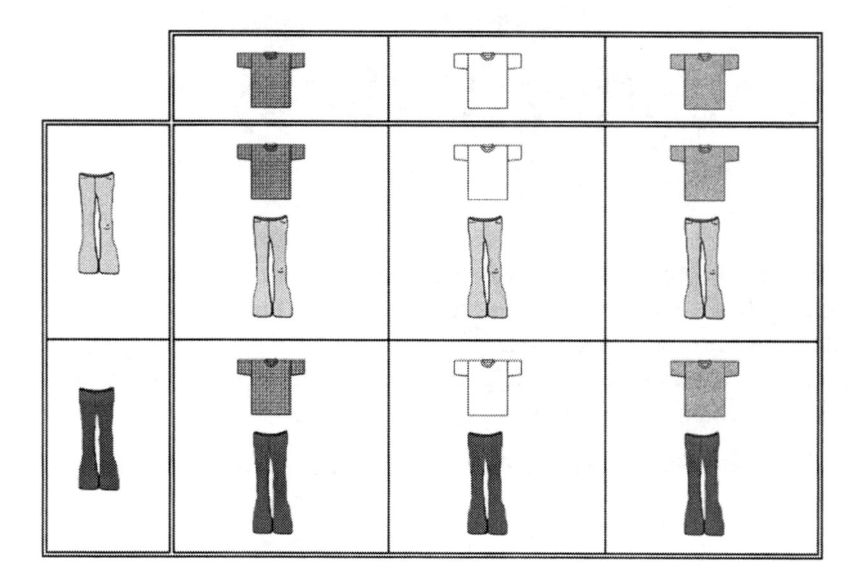

Figure 4.1
A grid showing all possible ways to combine three tops with two pairs of pants.

A **tree diagram** gives another way to see all the different possibilities. We start with a point, and then draw lines to each of the three different tops. Then from each top, we draw a line to each of the two different pants. Finally, we draw lines to each of the six different outfits, as shown in Figure 4.2. This shows that there must be 3×2, or 6 different outfits. We could have started with the pants, getting two branches from the original point, then three branches from each of these. The result would then be $2 \times 3 = 6$, which is of course the same.

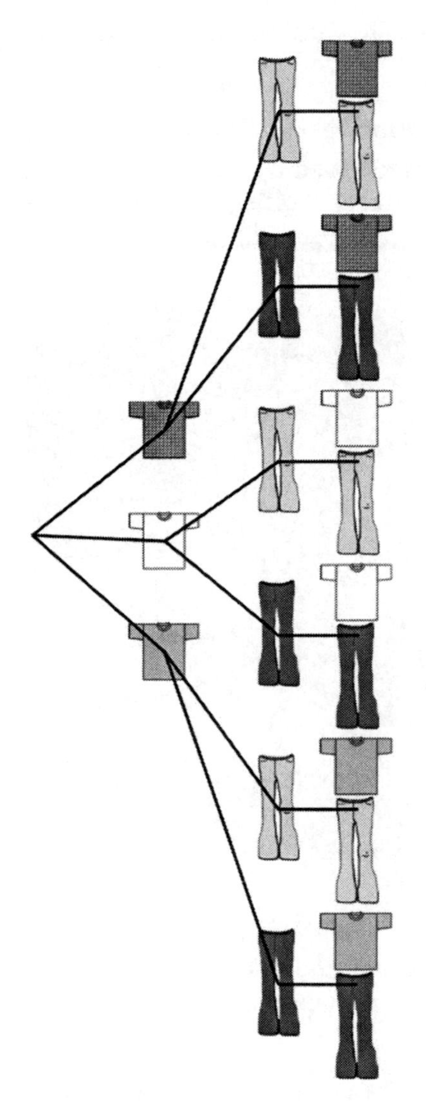

Figure 4.2
A tree diagram showing all possible ways to combine three tops with two pairs of pants.

Both of these pictures show us that if we make one choice in three ways and the other choice in two ways, we end up with 3×2, or 6 different choices. This example illustrates the basic counting principle for pairs.

NOTE

Product Rule for Counting Pairs (Counting Principle for Pairs)

If there are m choices for the first item followed by n choices for the second item, then there are mn possible choices of a pair.

Examples of the Product Rule for Counting Pairs

- When two dice are thrown, there are 36 possible outcomes, since there are six possible outcomes for each die. We count 1 on the first die and 6 on the second die as a different outcome from 6 on the first die and 1 on the second die.
- The Shanti Café offers a sundae made with one scoop of ice cream and one topping. The ice cream flavors are vanilla, chocolate, and coffee. The toppings are caramel, hot fudge, butterscotch, and strawberry. There are $3 \times 4 = 12$ possible sundaes.
- One senior and one junior are chosen to speak at a school pep rally. There are 50 seniors and 60 juniors in the school. By the Product Rule, there are $50 \times 60 = 3{,}000$ different possible ways to choose speakers.

Sometimes, we may make more than two choices—there may be more than two collections to choose from, or we may do an experiment more than twice. For example, you might wear a shirt, pants, and a sweater, or you might toss three coins.

If we toss three coins, we won't be able to draw a grid to show all possible outcomes, but we can draw a tree diagram, as shown in Figure 4.3. For each coin, there are two possibilities. We get three levels of branching and end up with $2 \times 2 \times 2 = 8$ possible outcomes. As we can see from the tree, HHT is different from HTH and THH and we have to count these as different outcomes.

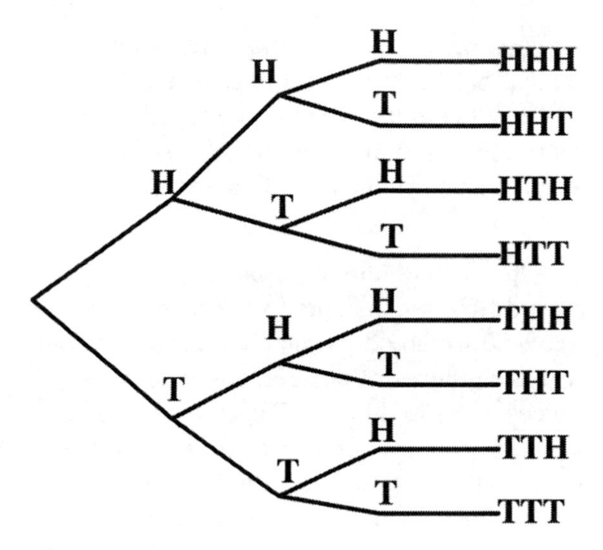

Figure 4.3
A tree diagram with three levels.

If it is more convenient, a tree diagram can be drawn so that the branches point downward, as shown in Figure 4.4.

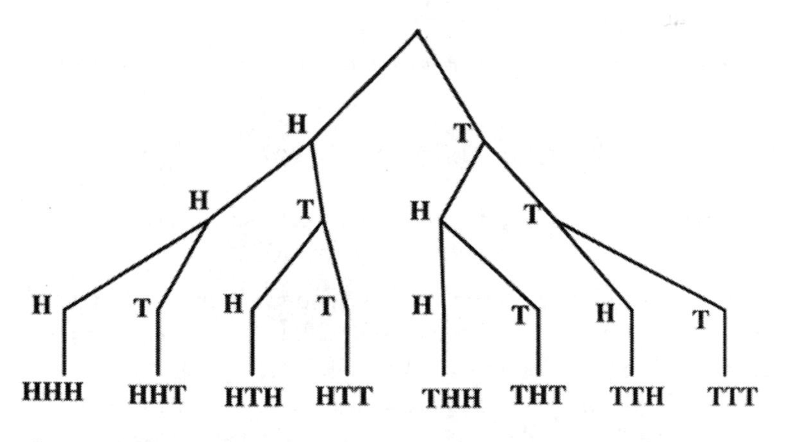

Figure 4.4
A tree diagram with three levels and branches pointing down.

We can extend this idea to any number of different choices and get the most general counting principle. That is, if we perform several different experiments or make many different choices, we multiply together the number of outcomes for each of the experiments or choices in order to get the total number of possible outcomes when all of the experiments or choices have been made.

For two choices, we used the letters m and n to represent how many possibilities there are. We use a different technique for more than two choices. Because we don't know how many different choices are possible, we use a variable i to represent the number of choices, and then use a subscript to tell us which choice we are looking at. Thus we write n_1 for the number of possible outcomes for the first choice, n_2 for the number of possible outcomes for the second choice, and so on until we get up to n_i for the number of possible outcomes for the ith, or last, choice.

NOTE

Product Rule for Counting (General Counting Principle)

If there are i choices made, with n_1 possibilities for the first choice, n_2 possibilities for the second choice, and so on, up to n_i, then there are $n_1 \times n_2 \times \ldots \times n_i$ possible outcomes.

Examples of the Product Rule for Counting

- When three dice are thrown, there are $6^3 = 216$ possible outcomes, since there are six possible outcomes for each die.
- The Cosmic Café offers a sundae made with one scoop of ice cream and one topping. The ice cream flavors are vanilla, chocolate, and coffee. The toppings are caramel, hot fudge, butterscotch, and strawberry. In addition, you can choose to have whipped cream; this gives two choices, with or without whipped cream. There are $3 \times 4 \times 2 = 24$ possible sundaes.

- One senior, one junior, and one sophomore are chosen to speak at a school assembly. There are 50 seniors, 60 juniors, and 80 sophomores in the school. By the Product Rule for Counting, there are $50 \times 60 \times 80 = 240,000$ different possible ways to choose speakers.
- Suppose there are r different balls and n different cells and we place each of the balls in one of the cells, allowing any number of balls to end up in any cell. Then there are r choices, one choice for each ball, with n possible outcomes, one for each cell. Here, we use the Product Counting Rule for the r different choices. Thus there are $n \times n \times \ldots \times n = n^r$ possible outcomes. This is a particularly important example because it can be used to model many different situations.

Sum Rule for Counting

Sometimes, you may be required to make one choice from two different groupings. To count up the number of choices in such a case, you will use addition. For example, a menu may list three different kinds of pizzas and four different pasta dishes. In this case, you have $3 + 4 = 7$ different choices.

It is important to note that if you are using addition to count choices, the collections of choices must be disjoint. For example, if a menu lists four different pasta dishes and six different entrees, but lists lasagna under both, we can't add 4 and 6 to get the number of different choices.

NOTE

Sum Rule for Counting

If you make one choice from two disjoint sets, one with m elements and one with n elements, then the number of choices you have is $m + n$.

Examples of the Sum Rule for Counting

- Madison has three different pairs of pants and two different pairs of shorts, so she has $3 + 2 = 5$ different possible choices of what to wear with her new sweater.
- If you can choose to roll one die, toss a coin, or draw one card from a deck of 52, there $6 + 2 + 52 = 70$ different possible outcomes.
- One speaker, who must be a senior or a junior, is to be chosen to speak at Convocation. There are 50 seniors and 60 juniors in the school. By the Sum Rule for Counting, there are $50 + 60 = 110$ different possible ways to choose a speaker.

4.2 Permutations

In this section and the next section, we will consider permutations and combinations, which are special applications of the Multiplication Rule. In a permutation, the order of the elements is important, but in a combination, the order is not important.

A **permutation** is an ordering or arrangement of a set of elements. For a set of two elements, {A, B}, there are two permutations or orderings of the elements: AB and BA. For a set of three elements, {A, B, C}, there are six permutations: ABC, ACB, BAC, BCA, CAB, and CBA.

Order is important when listing the winners of a competition, when performing tasks in sequence, or making a word out of letters. If each choice is given a different name or title, such as when officers are named for a club, then order is also important, and we use the technique of counting permutations.

To understand how to use the Multiplication Rule to count permutations, let's look at a set of four elements, {A, B, C, D}. One of the letters is chosen first, and there are four ways that can be done. When we choose the second letter, there are only three letters left to choose from. After the second letter in the ordering has been chosen, there are only two letters left to choose from, so there are two choices. The fourth or final letter can only be the one remaining letter, so there is only one possible choice. Thus the number of ways to permute, or order, four elements is $4 \times 3 \times 2 \times 1 = 24$. These are listed in the grid shown in Figure 4.5.

ABCD	BACD	CABD	DABC
ABDC	BADC	CADB	DACB
ACBD	BCAD	CBAD	DBAC
ACDB	BCDA	CBDA	DBCA
ADBC	BDAC	CDAB	DCAB
ADCB	BDCA	CDBA	DCBA

Figure 4.5
All permutations of four elements, A, B, C, and D.

In this case, we say that we have permuted four elements and use the notations ${}_4P_4$ or P(4,4), which both mean the number of permutations of four elements.

Since this type of calculation is common, there is a special notation and name for such expressions. The **factorial** of a positive integer is the product of that integer with all smaller positive integers. We write 4! for the product $4 \times 3 \times 2 \times 1$ and say "four factorial." For each positive integer n, we have $n! = n \times (n - 1) \times (n - 2) \times \ldots \times 3 \times 2 \times 1$ and say "n factorial." By convention, $0! = 1$. Most calculators have a factorial key to make computing factorials easy.

It may happen that only some of the elements are chosen in an arrangement. For example, if we want to choose only three of the five letters in the set {A, B, C, D, E} and arrange them in order, we again have five choices for the first letter, four choices for the second, and then three choices for the third and final letter. This gives us $5 \times 4 \times 3 = 60$ permutations, listed in the grid shown in Figure 4.6.

ABC	BAC	CAB	DAB	EAB
ABD	BAD	CAD	DAC	EAC
ABE	BAE	CAE	DAE	EAD
ACB	BCA	CBA	DBA	EBA
ACD	BCD	CBD	DBC	EBC
ACE	BCE	CBE	DBE	EBD
ADB	BDA	CDA	DCA	ECA
ADC	BDC	CDB	DCB	ECB
ADE	BDE	CDE	DCE	ECD
AEB	BEA	CEA	DEA	EDA
AEC	BEC	CEB	DEB	EDB
AED	BED	CED	DEC	EDC

Figure 4.6
All permutations of five elements A, B, C, D, and E, taken three at a time.

In this case, we say that we have permuted five elements taken three at a time and use the notations ${}_5P_3$ or P(5, 3), which both mean the number of permutations of five elements taken three at a time.

Since ${}_5P_3$ is the same as ${}_5P_5$ except that the last two factors 2×1 are left off, we can write ${}_5P_3$ as $5! / 2!$. The general formula is this:

$${}_nP_r = P(n, r) = n! / (n - r)!$$

Many calculators have a function key that will compute ${}_nP_r$.

Examples of Permutations

- Five children must be arranged in a row for a photograph. How many ways can this be done? Solution: There are five ways to choose the child who will be on the far left, then four ways to choose the child to the right of that one, then three ways to choose the next child, two ways to choose the next child, and finally one way to choose the child on the far right. This gives ${}_5P_5 = 5! = 5 \times 4 \times 3 \times 2 \times 1 = 120$ different ways to arrange the five children for the photo.

- A president, secretary, and treasurer are to be chosen from the eight members of the Flyers Frisbee Club. How many ways can this be done? Solution: There are eight ways to choose the president; once the president has been chosen, there are seven ways to choose the secretary, and once the secretary has been chosen, there are six ways to choose the treasurer. Thus there are then ${}_8P_3 = 8! / 5! = 8 \times 7 \times 6 = 336$ possible ways to choose the officers of the club.
- ${}_nP_0 = n! / n! = 1$
- ${}_nP_n = n! / 0! = n! / 1 = n!$

4.3 Combinations

A **combination** is a choice of elements where order is not important, unlike permutations, where order is important. Places where you might use combinations include selecting a committee, choosing toppings for a pizza, and getting a bridge hand.

For example, suppose a committee of three members is to be chosen from a club of six members. This is different from choosing a president, secretary, and treasurer because it does not matter in what order the members are chosen. If the members of the club are labeled A, B, C, D, E, and F, the possible committees are listed in Figure 4.7.

ABC	ACE	BCD	BEF
ABD	ACF	BCE	CDE
ABE	ADE	BCF	CDF
ABF	ADF	BDE	CEF
ACD	AEF	BDF	DEF

Figure 4.7
Combinations of three letters chosen from the six letters A, B, C, D, E, and F.

We use the Multiplication Rule to count the combinations of three letters chosen from the set of six letters {A, B, C, D, E, F}. Start the same way as with permutations, choosing one of the letters first; there are six ways that can be done. When we choose the second letter, there are only five possibilities, as one of the six letters has already been chosen. In the same way, when the third letter is chosen, there are only four elements left to choose from, so there are four choices. We get $6 \times 5 \times 4 = 120$ possibilities, but these are all permutations, and many combinations have been counted more than once. In fact, each triple, like ABC, has been counted six times, since there are six different permutations (or orderings) of three letters. To compensate for this, we divide 120 by 6, getting 20 combinations, and that is exactly how many were shown in Figure 4.7.

In this case, we say that we have chosen six elements from six elements and use the notations ${}_6C_3$ or C(6, 3), which both mean the number of combinations of three elements taken from a set of six elements. You can read this as "6 choose 3." Another notation for this is also used,

$$\binom{n}{r}$$

where n is the number of elements available and r is the number of elements chosen.

Since ${}_nC_r$ is the same as ${}_nP_r$ divided by $r!$, we can write the general formula for combinations this way:

$${}_nC_r = C(n, r) = \binom{n}{r} = n! \,/\, r!(n - r)!$$

Many calculators have a function key that will compute ${}_nC_r$.

Examples of Combinations

- Five children must be chosen from a class of 20 for a photograph. How many ways can this be done? Solution: There are 20 ways to choose the first child, 19 ways to choose the next child, 18 ways to choose the third child, 17 ways to choose the fourth child, and finally 16 ways to choose the fifth child. Since order doesn't matter, we must divide the product of these numbers by 5!, the number of ways to arrange the five children. This gives ${}_{20}C_5 = C(20, 5) = 20! / (5! \times 15!) = (20 \times 19 \times 18 \times 17 \times 16) / (5 \times 4 \times 3 \times 2 \times 1) = 15{,}504$ different ways to choose the five children for the photo.

- An executive committee of three members is to be chosen from the eight members of the Math Club. How many ways can this be done? Solution: There are eight ways to choose the first member of the committee, seven ways to choose the next member, and six ways to choose the third member. Since the order in which the members are chosen does not matter, we must divide this number by 3!. Thus there are then ${}_8C_3 = C(8, 3) = 8! / (3! \times 5!) = (8 \times 7 \times 6) / (3 \times 2 \times 1) = 56$ possible ways to choose the officers of the club.
- ${}_nC_0 = n! / (n! \times 0!) = n! / (n! \times 1) = 1$
- ${}_nC_n = n! / (0! \times n!) = n! / (n! \times 1) = 1$

4.4 Partitions

Sometimes we want to count up how many ways a set can be broken up into a collection of subsets. For example, a teacher may want to create four study groups of five students each from a class of 20 students.

One way to analyze this problem is to choose a group of five students from the whole class of 20, then five from the remaining 15 students, then five from the 10 students for the third group, leaving five students for the fourth group. From this, we see that there are ${}_{20}C_5 \times {}_{15}C_5 \times {}_{10}C_5 \times {}_5C_5$ ways to choose the four groups of five students. Calculating the numbers, we get ${}_{20}C_5 \times {}_{15}C_5 \times {}_{10}C_5 \times {}_5C_5 = 15{,}504 \times 3{,}003 \times 252 \times 1 = 11{,}732{,}745{,}024$.

To find the number of different partitions of a set, start by finding the number of different ways to form the first partition. From the remaining elements, find the number of different ways to form the second partition. Continue until all partitions are counted and multiply the resulting numbers, according to the Multiplication Rule for Counting.

Examples of Partitions

- Twelve different toys are to be distributed among four children, giving each child three toys. How many ways can this be done? Solution: There are ${}_{12}C_3 = 220$ ways to give toys to the first child, ${}_{9}C_3 = 84$ ways to give toys to the second child, ${}_{6}C_3 = 20$ ways to give toys to the third child, and one way to give the remaining toys to the fourth child. The product is $220 \times 84 \times 20 \times 1 = 369{,}600$.
- Three students each choose two problems from a list of eight problems, with no problem chosen more than once. How many ways can this be done? Solution: The first student has one of ${}_{8}C_2$ possible choices of problems. The second student then has one of ${}_{6}C_2$ choices and the third student has ${}_{4}C_2$ choices. There are ${}_{8}C_2 \times {}_{6}C_2 \times {}_{4}C_2 = 28 \times 15 \times 6 = 2{,}520$ possible ways to assign two problems to each of three students.
- From a list of 20 problems, a teacher wants to make up three homework assignments with six, five, and seven problems. How many ways can this be done? Solution: There are ${}_{20}C_6$ ways to choose the set of six problems, then ${}_{14}C_5$ ways to choose the set of five problems, and finally ${}_{9}C_7$ ways to choose the set of seven problems. There are ${}_{20}C_6 \times {}_{14}C_5 \times {}_{9}C_7 = 38{,}760 \times 2{,}002 \times 36 = 2{,}793{,}510{,}720$ possible ways to create the three homework assignments.
- Three students are studying together and have to complete an assignment of eight problems. They decide one student will do four problems, another student will do three problems, and the third student will do one problem. How many different ways can they distribute the problems? Solution: There are ${}_{8}C_4 \times {}_{4}C_3 \times {}_{1}C_1 = 70 \times 4 \times 1 = 280$ ways to split up the problems. Then there are $3! = 6$ ways to determine which student gets which number of problems. Thus there are $280 \times 6 = 1{,}680$ ways to do the assignment.

To get a general formula, suppose there are n objects that we wish to partition into i groups of sizes $r_1, r_2, r_3, \ldots,$ and r_i.

We can look at this problem another way when the sum $r_1 + r_2 + r_3 + \ldots + r_i = n$. Look at all possible permutations of all n items, and there are $n!$ of them. Since the first r_1 of the objects are together in a group where order doesn't matter, and this group has $r_1!$ possible arrangements, we must divide by $r_1!$. We do the same for each of the other groups. This gives an equivalent formulation of the Partition Rule for the number of ways to choose the groups.

NOTE

Partition Rule

The number of partitions of n objects into i groups of sizes r_1, r_2, r_3, . . . , and r_i, is

$$_{n}C_{r_1} \times {}_{n-r_1}C_{r_2} \times {}_{n-r_1-r_2}C_{r_3} \times \ldots \times {}_{n-r_1-r_1-\ldots-r_{i-1}}C_{r_i}$$

NOTE

Partition Rule

The number of partitions of n objects into i groups of sizes r_1, r_2, r_3, . . . , and r_i with $r_1 + r_2 + r_3 + \ldots + r_i = n$ is

$$\frac{n!}{r_1!r_2!\ldots r_i!}$$

In a case where the sum of the group sizes $r_1 + r_2 + r_3 + \ldots + r_i$ is not equal to n, you can add on the group of size $n - (r_1 + r_2 + r_3 + \ldots + r_i)$.

Examples of the Partition Rule

- Twelve different toys are to be distributed among four children. Each child gets three toys. Using the formula, we get

$$\frac{12!}{3!3!3!3!} = 369{,}600$$

ways to distribute the toys.

- Three students each choose two problems from a list of eight problems. How many ways can this be done? Since the three students don't choose all eight problems, we make a grouping for the two leftover problems. The formula gives $8! / (2! \times 2! \times 2! \times 2!) = 2{,}520$ ways to assign the problems.
- Three students are studying together and have to complete an assignment of eight problems. They decide one student will do four problems, another student will do three problems, and the third student will do one problem. The formula gives the number of ways to partition the problems: we multiply by $3! = 6$, the number of ways to assign the differently sized groups of problems to the three students, so we get $6 \times 8! / (4! \times 3! \times 1!) = 6 \times 280 = 1{,}680$ ways to do the assignment.

4.5 Challenging Counting Problems

There are many examples of challenging counting problems that use permutations and combinations along with the Sum Rule, the Multiplication Rule, or the Partition Rule, or they might involve dividing by a number that represents repeats in the counting. Each problem is different and must be solved by carefully considering the information given in the problem. Sometimes, drawing a tree diagram can be helpful.

Examples of Counting Problems

- How many passwords can you make if each password must consist of two letters, uppercase or lowercase, followed by four digits that must be different? Solution: There are 52 upper- and lowercase letters and 10 digits, so there are $52^2 \times {}_{10}P_4 = 2{,}704 \times 5{,}040 = 13{,}628{,}160$ different passwords.
- How many ways can six family members sit around a circular table? Assume that only the relative position of the family members matters, not which seat they are sitting in. Solution: The first person can sit anywhere. Then there are $5! = 120$ ways to arrange the other members. Thus there are 120 different seating arrangements.

- How many ways can six keys be put on a key chain? Ignore the shape of the keys; you are only interested in the ordering of the keys. Solution: There is only one way to put the first key on the chain. Then there are $5! = 120$ ways to put the next keys on the chain. However, of these 120, each arrangement and its mirror image look the same on a key chain and should not be counted twice. For example, ABCDEF and FEDCBA look the same on a key chain. Thus there are only $120 / 2 = 60$ different ways to arrange the keys on the chain.
- How many different monograms are there with one initial for the last name (surname) and one or two initials for the given names? Solution: For this problem, we use both the Multiplication Rule and the Sum Rule. There are $26 \times 26 = 676$ monograms with two initials and $26 \times 26 \times 26 = 17{,}576$ monograms with three initials, so there are a total of $676 + 17{,}576 = 18{,}252$ monograms consisting of two or three initials.
- How many different six-letter "words" can be written with the letters of the word *passes*? Solution: There are six letters in the word *passes*, so there are ${}_6P_6 = 6!$ six-letter words that use the letters of *passes*. However, the letter *s* occurs three times; rearranging these won't give a different word, so we must divide by 3! and get $6! / 3! = 720 / 6 = 120$ different words.

4.6 Binomial Coefficients

A **binomial term** is the sum or difference of two numbers or expressions, like $a + b$, $x + 3$, or $b - 7$. In algebra, we often encounter powers of binomial terms, expressions like $(a + b)^2$ or $(a + b)^3$ that we need to expand. We expand $(a + b)^3$ by performing the following steps:

$$\begin{aligned}(a + b)^3 &= (a + b)^2 \times (a + b)\\ &= (a^2 + 2ab + b^2) \times (a + b)\\ &= a^3 + 3a^2b + 3ab^2 + b^3\end{aligned}$$

Binomial coefficients are the numbers or coefficients that are in front of the variable terms, like a^3 and a^2b in the expansion of an expression like $(a + b)^n$. For example, the binomial coefficient of the term a^2b in the expansion of $(a + b)^3$ is 3.

Using our knowledge of combinations, we can do this kind of computation easily. We look at the factors written out and see how the different terms in the answer are obtained. So for the example of $(a + b)^3$, we get three factors.

$$(a + b)^3 = (a + b) \times (a + b) \times (a + b)$$

How can we get the term a^3 in the final answer? There is only one way, selecting the term a in each of the factors; this means that the coefficient of a^3 is 1. For the term a^2b, we choose a from two of the three factors and b from the other factor; and there are ${}_3C_2 = 3$ ways to do this. For the next term, ab^2, we choose a from one of the three factors and b from the other two; there are ${}_3C_1 = 3$ ways to do this. For the final term, we don't choose any a terms and there is ${}_3C_0 = 1$ way to do that. Putting these observations together we get $a^3 + 3a^2b + 3ab^2 + b^3$.

This reasoning will work for every positive power n, so we get the **Binomial Theorem**:

$$(a + b)^n = {}_nC_n a^n + {}_nC_{n-1} a^{n-1}b + {}_nC_{n-2} a^{n-2}b^2 + {}_nC_{n-3} a^{n-3}b^3 + \ldots + {}_nC_1 ab^{n-1} + {}_nC_0 b^n$$

Examples of Binomial Coefficients

- $(x + y)^6 = {}_6C_6x^6 + {}_6C_5x^5y + {}_6C_4x^4y^2 + {}_6C_3x^3y^3 + {}_6C_2x^2y^4 + {}_6C_1xy^5 + {}_6C_0y^6$
  $= x^6 + 6x^5y + 15x^4y^2 + 20x^3y^3 + 15x^2y^4 + 6xy^5 + y^6$
- In the expansion of $(a + b)^{10}$, the coefficient of the term a^6b^4 is ${}_{10}C_6 = 10! / (6! \times 4!) = 210$.

4.7 Pascal's Triangle

Pascal's triangle is a special arrangement of the binomial coefficients discussed in the previous section. It gives an easy way to display and remember the binomial coefficients, and it has many interesting properties.

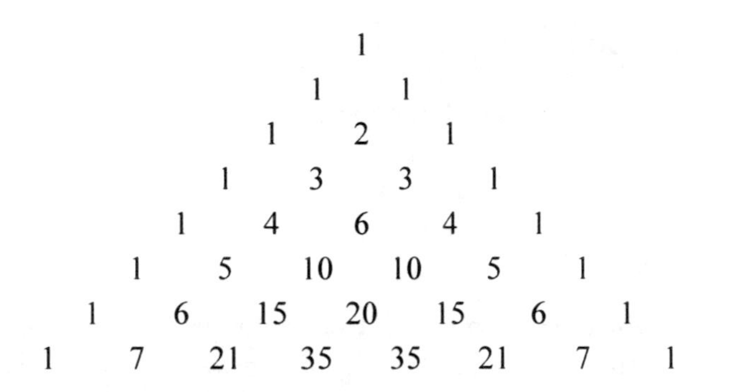

The number at the top and the numbers at the edges are always 1. Every other number is the sum of the two numbers above it, one slightly to the left and one slightly to the right. For example, the number 20 is the sum of the two numbers above it, $20 = 10 + 10$.

If we start counting rows with the row containing the numbers 1 and 1, we see that the *n*th row gives the binomial coefficients of the expansion for $(a + b)^n$. The coefficient of $a^i b^{n-i}$ is ${}_nC_i$ and is the $(n - i + 1)$th number in the *n*th row. It is conventional to call the top row the 0th row. Recall the rule from algebra that every number raised to the 0 power is 1—and now this row also follows the pattern since $(a + b)^0 = 1$.

Since each number in Pascal's triangle is the sum of the two numbers above it, and all three of these numbers can be expressed as combinations, we get this important relationship:

$$\binom{n+1}{r} = \binom{n}{r-1} + \binom{n}{r}$$

Examples of Using Pascal's Triangle

- $(x + y)^5 = x^5 + 5x^4y + 10x^3y^2 + 10x^2y^3 + 5xy^4 + y^5$ (We use the fifth row of Pascal's triangle to get the coefficients.)
- The coefficient of a^5b^2 in $(a + b)^7$ is 21. This is the third number in the seventh row of Pascal's triangle.

Chapter 4 Summary

A **tree diagram** organizes different possibilities or outcomes to make it easier to count them. Starting with a point, draw lines to each of the first possible outcomes. Then from each of these, draw lines to each of the next possible outcomes. Continue until all possible stages have been taken into account.

Product Rule for Pairs (Counting Principle for Pairs). If there are m choices for the first place and n choices for the second place, there are mn possible pairs.

Product Rule for Counting (General Counting Principle). If there are i choices made, with n_1 possibilities for the first choice, n_2 choices for the second place, and so on, up to n_i, then there are $n_1 \times n_2 \times \ldots \times n_i$ possible choices.

Sum Rule for Counting. If you make one choice from two disjoint sets, one set with m elements and one set with n elements, then the number of different possible choices is $m + n$.

The **factorial** of a positive integer is the product $n! = n \times (n - 1) \times (n - 2) \times \ldots \times 3 \times 2 \times 1$ and say "n factorial." By convention, $0! = 1$.

A **permutation** is an ordering or arrangement of elements taken from a set.

The number of permutations of n objects taken r at a time is as follows:

$${}_nP_r = P(n, r) = n! / (n - r)!$$

$${}_nP_0 = n! / n! = 1$$

$${}_nP_n = n! / 0! = n! / 1 = n!$$

A **combination** is a choice of elements from a set; the order of the elements is not important.

The number of combinations of n objects taken r at a time is

$${}_nC_r = C(n, r) = \binom{n}{r} = n! / r!(n - r)!$$

$${}_nC_0 = n! / (n! \times 0!) = n! / (n! \times 1) = 1$$

$${}_nC_n = n! / (0! \times n!) = n! / (1 \times n!) = 1$$

Partition Rule. The number of ways to partition n objects into a collection of sets having $r_1, r_2, r_3, \ldots, r_i$ elements where $r_1 + r_2 + r_3 + \ldots + r_i = n$ is

$$\frac{n!}{r_1!r_2!\ldots r_i!}$$

A **binomial term** is the sum or difference of two numbers or expressions, like $a + b$ or $x + 3$.

Binomial coefficients are the numbers or coefficients that are in front of the variable terms like a^3 and a^2b in the expansion of an expression like $(a + b)^n$.

Binomial Theorem. The expansion of a binomial term raised to the power n is

$$(a + b)^n = {}_nC_n a^n + {}_nC_{n-1}a^{n-1}b + {}_nC_{n-2}a^{n-2}b^2 + {}_nC_{n-3}a^{n-3}b^3 + \ldots + {}_nC_1 ab^{n-1} + {}_nC_0 b^n.$$

Pascal's Triangle is a special arrangement of the binomial coefficients.

1
1 1
1 2 1
1 3 3 1
1 4 6 4 1
1 5 10 10 5 1
1 6 15 20 15 6 1
1 7 21 35 35 21 7 1

The number at the top and the numbers at the edges are always 1. Every other number is the sum of the two numbers above it, one slightly to the left and one slightly to the right.

The row containing 1 1 is counted as the first row, and the nth row gives the binomial coefficients of the expansion for $(a + b)^n$. The coefficient of a^ib^{n-i} is ${}_nC_i$ and is the $(n - i + 1)$th number in the nth row.

The top row is the 0th row and $(a + b)^0 = 1$.

For all binomial coefficients,

$$\binom{n+1}{r} = \binom{n}{r-1} + \binom{n}{r}$$

Chapter 4 Practice Problems

1. Aruna has four blouses, three sweaters, and five skirts. How many outfits does she have consisting of a blouse and a skirt? How many outfits consist of a blouse, a skirt, and a sweater? Which counting rule did you use?
2. Draw a grid and tree diagram to show all possible outcomes when you toss two coins. How many outcomes are there?
3. Driving down Main Street, you go through three intersections where there are traffic lights. Each light can be green, red, or yellow. Draw a tree diagram to show all possible ways you can encounter the traffic lights. How many different sequences of lights could you encounter?
4. Some games are played with tetrahedral dice, which have four sides and show the numbers 1, 2, 3, or 4 when thrown. If you toss three tetrahedral dice, how many outcomes are possible?
5. At the upscale gourmet restaurant Chateau Bois Aubry, there are three choices for appetizer, two for soup, five for entrée, three for salad, and eight for dessert. How many different meals are possible at the restaurant, assuming each meal consists of an appetizer, soup, entrée, salad, and dessert? Which counting principle did you use?
6. Sandi has three pairs of sandals, four different pairs of athletic shoes, and two pairs of boots. How many choices does she have for footwear each morning?
7. How many different five letter "words" can be made out of the letters in the word *world*?
8. Five friends want to create a video together. One will be the scriptwriter, one will be the director, and the others will act in the video. How many ways are there to choose the scriptwriter and director? List all of the possibilities, assuming that the friends are labeled A, B, C, D, and E.

9. Twelve soccer teams compete in a tournament. How many different possible outcomes for first place, second place, and third place are there?
10. The manager of the winning soccer team chooses three of the 20 members to be interviewed by a sports reporter. How many ways can they be chosen?
11. For each of the following, you are interested in how many ways of choosing different outcomes there are. Tell whether it is a situation involving permutations or combinations. In other words, is the order in which the objects are selected important or not?
    - **a.** The winning lottery numbers in Powerball
    - **b.** The numbers called out in a game of Bingo
    - **c.** The digits in a number
    - **d.** The letters in a word
    - **e.** The birthdays of 20 people
    - **f.** Books on a shelf
    - **g.** Cards in a poker hand
    - **h.** Misprints on a page of a book
    - **i.** Defective products from an assembly line
    - **j.** People sitting around a table
    - **k.** Letters and digits on a license plate
    - **l.** A committee chosen from a political party
    - **m.** Officers elected by a club
    - **n.** Choosing 10 questions to answer on an exam
12. How many ways can two keys be put on a keychain? Three keys? Four keys?
13. How many different monograms are there with one initial for the surname and at most three initials for the given names?
14. How many different 10-letter "words" can be made from the letters in the word *invincible*?
15. Write out the two lines of Pascal's triangle following the last line given in the text. The first of these lines should begin with the numbers 1 and 8.
16. What is ${}_8C_5$?
17. Expand $(a + b)^9$.
18. What is the coefficient of the term x^4y^4 in the expansion of $(x + y)^8$?

Answers to Chapter 4 Practice Problems

1. There are 20 outfits consisting of a skirt and blouse; 60 outfits consisting of a skirt, blouse, and sweater. Use the Product Rule for Counting.
2. There are four outcomes. See Figures 4.8 and 4.9.

	H	T
H	HH	HT
T	TH	TT

Figure 4.8
A grid showing all possible outcomes when tossing two coins.

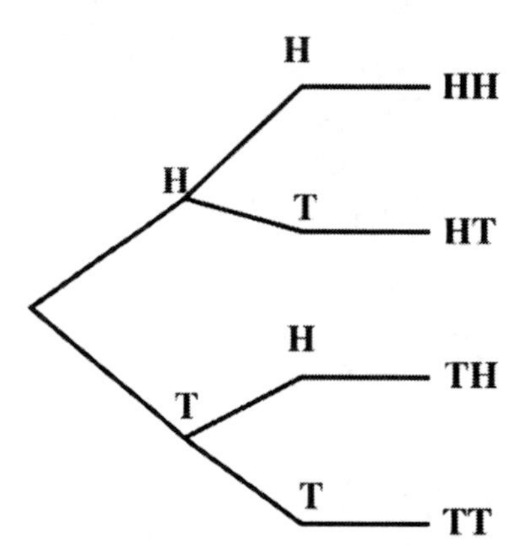

Figure 4.9
A tree showing all possible outcomes when tossing two coins.

3. Since the problem asks for sequences of lights, order is important, and we must use permutations. There are 27 possible different sequences of traffic lights. See Figure 4.10.

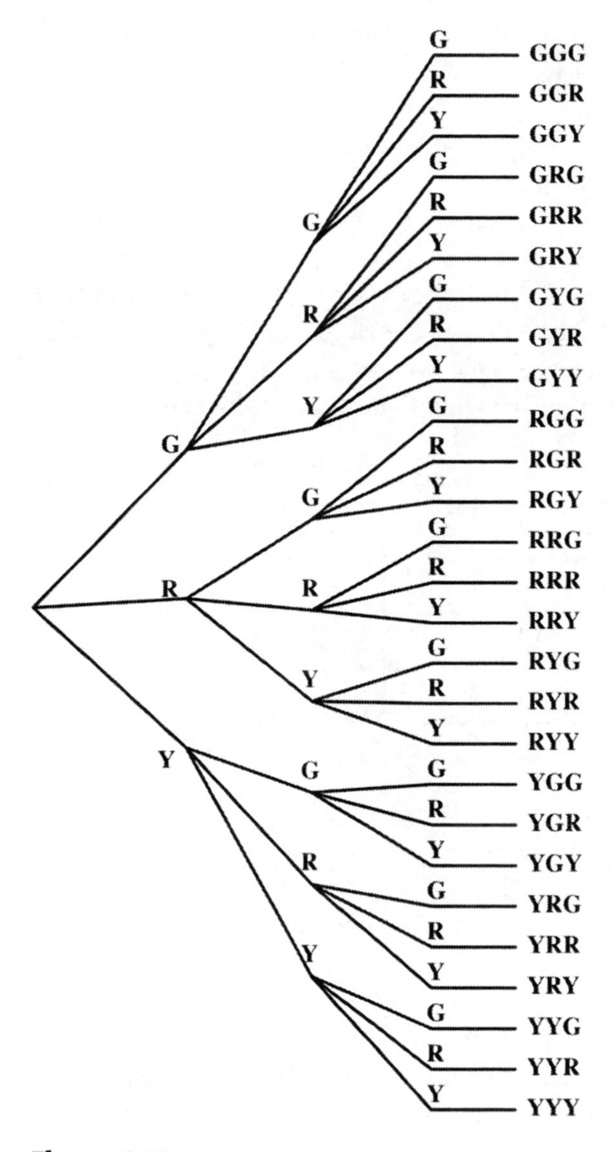

Figure 4.10

A tree showing all 27 sequences of traffic lights when traveling down Main Street.

4. We must use permutations since, for example, throwing 1,1,2 is a different outcome from 1,2,2. There are $4 \times 4 \times 4 = 64$ different outcomes.
5. There are $3 \times 2 \times 5 \times 3 \times 8 = 720$ possible meals. Use the Product Rule for Counting.
6. Sandi has $3 + 4 + 2 = 9$ choices of footwear. Use the Sum Rule for Counting.
7. There are ${}_5P_5 = 5! = 5 \times 4 \times 3 \times 2 \times 1 = 120$ different "words" that can be made from the letters in the word *world*.
8. There are ${}_5P_2 = 5 \times 4 = 20$ ways to choose scriptwriter and director. They are given in the grid shown in Figure 4.11, with scriptwriter listed first.

AB	BA	CA	DA	EA
AC	BC	CB	DB	EB
AD	BD	CD	DC	EC
AE	BE	CE	DE	ED

Figure 4.11
A grid showing all possible ways to choose scriptwriter and director.

9. There are ${}_{12}P_3 = 12 \times 11 \times 10 = 1{,}320$ possible outcomes for first, second, and third.
10. There are ${}_{20}C_3 = (20 \times 19 \times 18) / (3 \times 2 \times 1) = 1{,}140$ ways to choose three team members to be interviewed.
11. **a.** Combination. (It doesn't matter in what order the winning numbers are drawn, only that the numbers are the same as on the lottery ticket.)
    b. Permutation (a different order in calling the numbers can result in a different winner)
    c. Permutation
    d. Permutation
    e. Combination. (You are just interested in where the birthdays fall,

not in the order in which the 20 people are listed.)

f. Permutation

g. Combination

h. Combination. (You are just interested in which letters are printed incorrectly, not in the order you find them.)

i. Combination. (You are just interested in how many products are defective, not in what order you find them.)

j. Permutation

k. Permutation

l. Combination

m. Permutation

n. Combination

12. For two keys, one way: AB; for three keys, one way: ABC; for four keys, three ways: ABCD, ACBD, ACDB.

13. There are $26 \times 26 = 676$ monograms with two initials, $26 \times 26 \times 26 = 17{,}576$ monograms with three initials, and $26 \times 26 \times 26 \times 26 = 456{,}976$ monograms with four initials. Adding these together, we get 475,228 possible different monograms with up to four initials.

14. There are three copies of the letter i and two copies of the letter n, so there are $\frac{10!}{3!2!} = 302{,}400$ different possible "words."

15.

1 8 28 56 70 56 28 8 1

1 9 36 84 126 126 84 36 9 1

16. ${}_8C_5 = \frac{8!}{5!3!} = 56$ (This is the sixth entry in the eighth row of Pascal's triangle.)

17. $(a + b)^9 = a^9 + 9a^8b + 36a^7b^2 + 84a^6b^3 + 126a^5b^4 + 126a^4b^5 + 84a^3b^6 + 36a^2b^7 + 9ab^8 + b^9$

18. 70 (This is the fifth term in the eighth row of Pascal's triangle.)

Chapter 5

Computing Probabilities

In this chapter, you will learn how to compute probabilities in different situations from the definition of probability using the properties of sets from Chapter 3 and the counting techniques from Chapter 4. The examples and practice problems will help you master techniques which you will be applying over and over in all areas of probability

Recall that a **sample space** is the set of all possible outcomes of an experiment or observation; for now, assume that all outcomes are equally likely. An **event** is a specific subset of the sample space. A **trial** is one performance of the experiment or observation.

The probability of an event, when each element in the sample space is equally likely to occur, is as follows:

$$\frac{\text{number of elements in the event}}{\text{number of elements in the sample space}}$$

To use this formula, start by counting the elements in the sample space. Next, look at the description of the event and count how many elements of the sample space are in the event. Finally, compute the ratio.

You can solve many problems in probability in more than one way. Try solving some of the problems at the end of this chapter using different methods and see which you prefer.

5.1 About Numerical Computations

This section gives guidelines about the kinds of numbers and computations that appear frequently in probability. Many of the computations involve very small decimal fractions or very large numbers and are easier and faster with a calculator or computer.

Most scientific calculators have special function keys for factorials, permutations, and combinations. Calculators usually round very large or very small numbers and give them in scientific notation, described later in this section. If you have a scientific calculator, spend some time learning how to use the function keys, memory, and other special features.

Computer algebra systems like *Mathematica* or *Maple* give accurate answers for most integer computations, including factorials, permutations, and combinations. For decimal fractions, you can input the number of significant digits you want in the answer.

The **significant digits** of a number are all the digits in the number except 0s at the beginning or at the end of the number. The numbers 0.00057 and 5,700,000 both have two significant digits, since we don't count the 0s at the beginning or the end. The numbers 0.00507 and 507,000 both have three significant digits because in both cases the 0 between the 5 and the 7 is counted as a significant digit.

Examples of Significant Digits

The following numbers all have three significant digits.

- 143
- 104
- 1,030,000
- 0.138
- 0.000439

There are some exceptions to this rule. The ending zeros of an integer with a decimal point (like 1,030,000.) are significant digits. If a decimal fraction is given with additional zeros, they are significant, so 4.0030 has five significant digits.

Physical measurements always have some inaccuracies introduced by the nature of the measurement process. This inaccuracy is called **error of measurement**, and only the first few digits in a physical measurement are accurate. Often, an estimation of error will be given. For example, if we read that the distance to the moon is 239,000 ± 1,000 miles, then the digits 2, 3, and 9 are significant, and the true distance is in the range from 238,000 to 240,000.

Examples of Significant Digits That Are Zeros

The following numbers all have three significant digits.

- 140
- 100
- 0.100
- 14.0
- 0.130
- 0.000430

Many probabilities are decimals that go on forever. For example, the probability of getting a 1 or 2 when rolling one die is 2 / 6 = 1 / 3 = 0.33333. . . . For convenience, we round such numbers and retain only the digits that we want to use later on.

Rounding reduces the number of significant digits and leads to a small error in the last digit. Computing with numbers that have been rounded leads to a larger error in the last digits of the answer. For this reason, after a computation is done with numbers that have been rounded, we round the answer so that it has the same number of significant digits as the numbers that we have started with. For example, we multiply 0.333

and 0.467, which each have three significant digits, and get 0.155511. We round this number to 0.156, which also has three significant digits.

Rounding decimal numbers to two or three decimal places will give an answer that is easy to use and is accurate enough for most situations in a probability course.

To round a number, look at the digit after the last decimal place that you want to keep. So if you want to round 2.71895 to two decimal places, look at the thousandths digit, which in this case is 8. Since this digit is 5 or greater, add one to the hundredths digit and eliminate the thousandth digit and all digits to the right. So rounding 2.71895 to two decimal places gives 2.72 and rounding 4.78541 to two decimal places gives 4.79. If the digit after the last decimal place that you want to keep is less than 5, eliminate it and all digits to the right. So when we round 2.71443 to two decimal places, we get 2.71.

Rounding decreases the number of significant digits and also decreases the accuracy of computations with the rounded numbers. To minimize the decrease in accuracy, round only at the end of a series of computations.

Examples of Rounding Decimals

The numbers below have been rounded to two decimal places.

- $12.31998 \approx 12.32$
- $5.43128 \approx 5.43$
- $4.505 \approx 4.51$
- $3.999 \approx 4.00$
- $65.8319 \approx 65.83$

Scientific notation is useful for very large or very small numbers. We write a number as a product of a number between 0 and 10 (or between 0 and -10 if the number is negative) times a power of 10. The exponent is positive for large numbers and negative for small numbers. Often a number will be rounded before being written in scientific notation.

Examples of Writing Numbers in Scientific Notation

The numbers below have been re-written using scientific notation.

- 1,271,895 $\approx 1.27 \times 10^6$
- 0.00001271895 $\approx 1.27 \times 10^{-5}$
- 1,000,000 $= 10^6$

5.2 General Principles for Computing Probabilities

The general principles and techniques in this section will help you compute probabilities in many different situations. Examples show how to use each principle or technique.

Counting Events

When all possible outcomes of a trial are equally likely, we compute the probability of an event by using this formula:

$$\frac{\text{number of elements in the event}}{\text{number of elements in the sample space}}$$

If the sample space is small, we can list all of the elements in the sample space and count them, and then list all of the elements in the event and count them. Sometimes the sample space may be too large to count, but if we already know how many elements there are in the sample space, we need to list and count only the elements in the event.

Examples of Computing Probabilities

- Choose a number at random from the primes less than 20. What is the probability that it has two digits? Solution: The primes less than 20 are 2, 3, 5, 7, 11, 13, 17, and 19. There are eight primes and four have two digits. The probability is 4 / 8 = 0.5 that a number chosen at random has two digits.

- A month is chosen at random. What is the probability that the month begins with the letter *J*? Solution: There are 12 months, and January, June, and July begin with the letter *J*, so the probability of choosing a month beginning with *J* is 3 / 12 = 0.25.
- A card is chosen at random from a standard deck. What is the probability that it is a red face card? Solution: A standard deck has 52 cards. There are six red face cards, K♥, Q♥, J♥, K♦, Q♦, and J♦. The probability of a red face card is 6 / 52 ≈ 0.115.
- One of the 50 states is chosen at random. What is the probability that the name of the state has five or fewer letters? Solution: There are six states with names having five or fewer letters: Idaho, Iowa, Maine, Ohio, Texas, and Utah. The probability of getting one of these at random is 6 / 50 = 0.12.

Multiplication Rule for Probability

Sometimes we make a sequence of trials (of the same or different experiments or observations) and want to know about a certain sequence of outcomes. For example, we throw two dice four times in a row and wonder what the probability is of getting doubles on all four throws. Or we throw two dice, and then choose one card from a standard deck of 52 and want to know the probability of doubles followed by a face card. In these situations, we multiply the probabilities of success for each trial in the sequence to get the probability of success of the specific sequence of outcomes that we want. This is the Multiplication Rule for Probability for two trials. The rule for more trials is similar. There are some exceptions to this rule, such as when the outcome of one trial influences the outcome of another trial. These exceptions (when trials are not independent) will be covered in Chapter 6.

NOTE

Multiplication Rule for Probability

If we perform two trials, and the probability of success in the first trial is p and the probability of success in the second trial is q, then the probability of successes in both trials is pq.

Tree diagrams illustrate this principle. For example, you draw a marble from an urn that has two red and three blue marbles, and then a marble from another urn that has three green and seven yellow marbles. What are the probabilities of the different outcomes? Figure 5.1 shows all possible outcomes along with their probabilities in a tree diagram.

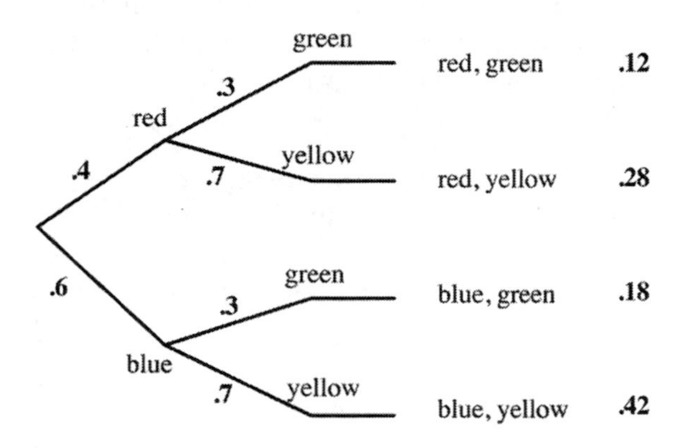

Figure 5.1
A tree diagram showing outcomes and probabilities.

The probability of a red marble and a green marble is 0.12; of a red and a yellow, 0.28; of a blue and a green, 0.18; and of a blue and a yellow, it is 0.42.

Examples of the Multiplication Principle

- Roll two dice four times. What is the probability of doubles four times in a row? Solution: The probability of doubles is $6 / 36 = 1 / 6 \approx 0.16667$, and the probability of doubles four times in a row is $(0.16667)^4 \approx 0.000772$, or 7.72×10^{-4} in scientific notation.
- If you roll two dice and then draw a card from a standard deck, what is the probability that you will get doubles followed by a face card? Solution: The probability of doubles is $6 / 36 = 1 / 6 \approx 0.16667$, and the probability of a face card is $12 / 52 \approx 0.23077$. The probability of doubles followed by a face card is $0.16667 \times 0.23077 \approx 0.0385$.

Complement Rule for Probability

Sometimes it is easier to find the probability that an event will *not* happen than it is to find the probability that it *will* happen. For example, when you toss four dice, it is easier to compute the probability that all four dice have different values than the probability of at least one pair. When you compute the probability that at least two dice will show the same number, you have to take into account the probabilities of two dice with the same number, three dice with the same number, and four dice with the same number, as well as the probability of two pairs.

An event and its complement make up the whole sample space, so the probability of an event added to the probability of its complement is 1.

> **NOTE**
> **Complement Rule**
> $P(A^C) = 1 - P(A)$

The Venn diagram in Figure 5.2 helps to explain why this is true. Every element of the sample space is either in A or in its complement A^C. The probability of the event that includes the whole sample is 1, so $P(A) + P(A^C) = 1$. When we subtract $P(A)$ from both sides, we get the complement rule.

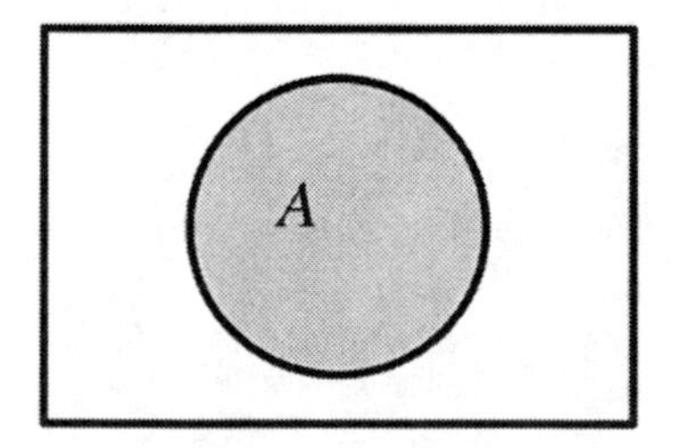

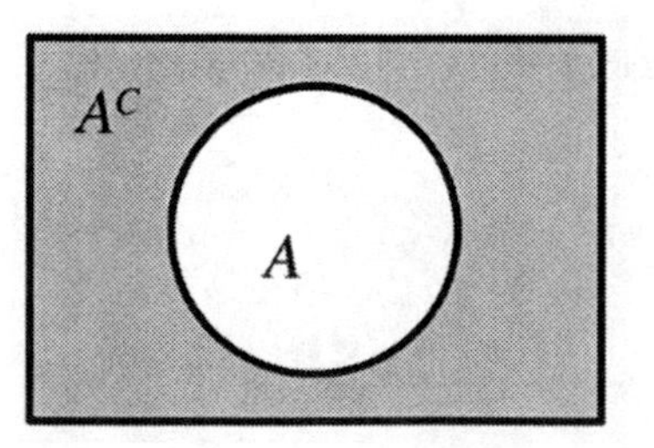

Figure 5.2
Venn diagrams of a set A and its complement A^C.

Example of the Complement Principle

If you roll four dice, what is the probability that there will be at least two dice showing the same number? Solution: It is easier to find the probability that the four numbers are all different. The first die can be any one of six numbers, the next die can be one of the other five numbers, the third die any one of four numbers, and the fourth die any one of the remaining three numbers, so there are $6 \times 5 \times 4 \times 3$ ways to get all different numbers on the four dice. There are 6^4 different outcomes for four dice. So we get $(6 \times 5 \times 4 \times 3) / 6^4 = 360 / 1296 \approx 0.278$ for the probability for all different numbers, and we get $1 - 0.278 \approx 0.722$ for the probability that there will be at least one pair when four dice are rolled.

Addition Rule of Probability for Mutually Exclusive Events

Sometimes we want to compute the probability of an event that is made up of two different events. For example, when you toss two dice, the event "doubles or an odd sum" is easier to work with if we think of the two separate events, "doubles" and "an odd sum." These two events are mutually exclusive; in other words, there is no outcome or element that is in both of the events.

NOTE

Addition Rule of Probability for Mutually Exclusive Events

If A and B are mutually exclusive events in the same sample space, then $P(A \cup B) = P(A) + P(B)$.

Look at the Venn diagram in Figure 5.3, which shows two disjoint sets. The number of events in $A \cup B$ is the sum of the number of events in A plus the number of events in B.

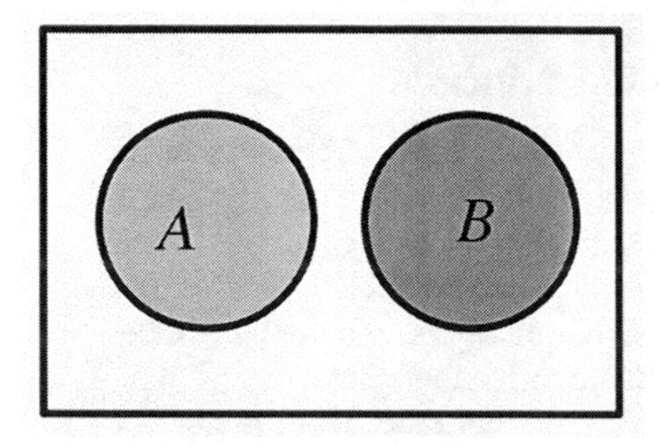

Figure 5.3
Two mutually exclusive events A and B, represented by disjoint sets.

Example of Mutually Exclusive Events

If you roll two dice, what is the probability that there will be doubles or an odd sum? Solution: Doubles, or both dice showing the same number, will always give an even sum, so doubles can never happen along with an odd sum—doubles and an odd sum are mutually exclusive. There are 6×6 outcomes for rolling two dice and six ways of getting a double, so $6 / 36 = 1 / 6 \approx 0.167$ is the probability of getting a double. To get an odd sum, you must have an odd number along with an even number. For each number on the first die, there are three possibilities for the second die to make an odd sum. This gives 6×3 ways to get an odd sum, with probability $18 / 36 = 1 / 2 = 0.5$. Finally, for the event of doubles or odd sum, we add these two probabilities to get $0.167 + 0.5 \approx 0.667$ for the answer.

We also could have solved this problem by counting the number of elements in the event. There are six doubles and 18 ways to get an odd sum, so there are $6 + 18 = 24$ elements, and the probability of doubles or an odd sum is $24 / 36 \approx 0.667$.

General Addition Rule for Probability

Maybe we want the probability of an event that is made up of two different events, but the events are not mutually exclusive. For example, when you toss two dice, the event "doubles or sum greater than 8" is easier to work with if we think of the two separate events, "doubles" and "sum greater than 8." These events are not mutually exclusive, since sometimes doubles gives a sum greater than 8.

> **NOTE**
>
> **General Addition Rule for Probability**
>
> If A and B are events in the same sample space, then $P(A \cup B) = P(A) + P(B) - P(A \cap B)$.

The Venn diagram in Figure 5.4 shows two sets A and B that intersect. The sum of the number of elements in A added to the number of elements in B counts the elements in the intersection $A \cap B$ twice. Subtracting off the number of elements in $A \cap B$ gives the number of elements in the union $A \cup B$.

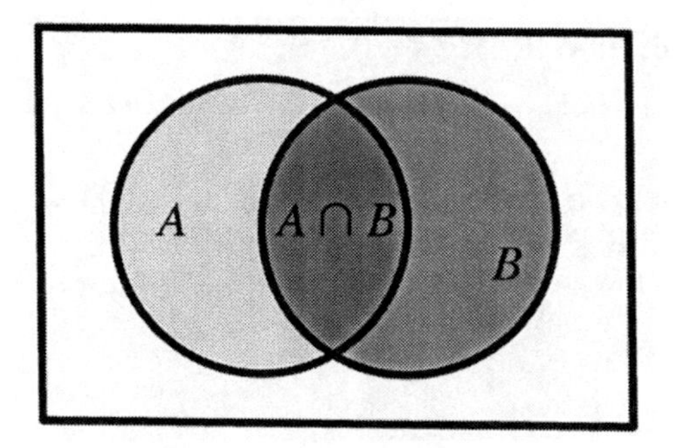

Figure 5.4
Venn diagram for two events A and B.

Example of the General Addition Rule

If you roll two dice, what is the probability that there will be doubles or a sum greater than 8? Solution: These two events are not mutually exclusive. There are 6 × 6 outcomes and six ways of getting doubles, so 6 / 36 = 0.167 is the probability of getting doubles. Listing the possible outcomes that give a sum greater than 8, we get the following: (3, 6), (4, 5), (4, 6), (5, 4), (5, 5), (5, 6), (6, 3), (6, 4), (6, 5), and (6, 6). There are 10 of these, so $10/36 \approx 0.278$ is the probability of a sum greater than 8. Two of these are doubles, with probability $2/36 \approx 0.056$ for the event doubles *and* sum greater than 8. Finally, for the event of doubles *or* sum greater than 8, we get that the probability is $0.167 + 0.278 - 0.056 = 0.389$.

Inequality for Probabilities

A subset has the same or fewer elements than the set containing it. For probabilities, this fact gives us an inequality.

NOTE

Probability Inequality

If event *A* is a subset of event *B*, then $P(A) \leq P(B)$.

The Venn diagram in Figure 5.5 helps us understand this inequality. This rule can help you estimate probabilities and sometimes identify errors.

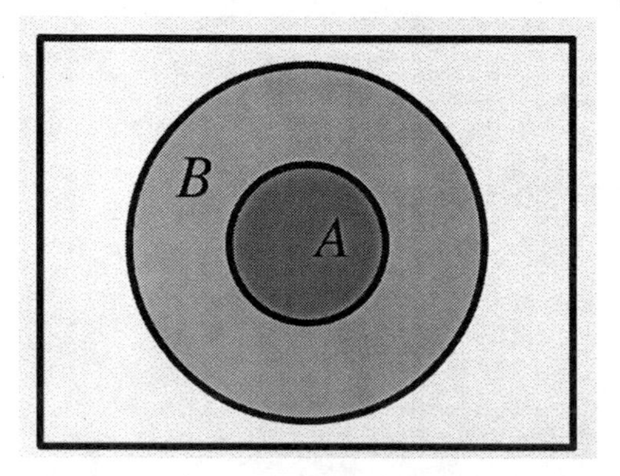

Figure 5.5
Venn diagram of event *A* that is a subset of event *B*.

Examples of Probability Inequalities

- You know that the probability of drawing a diamond from a deck is 0.25. If someone tells you the probability of drawing a diamond face card is 0.3, what do you say? If they say the probability of drawing an even numbered diamond is 0.15, what do you say? Solution: Since it is less likely to draw a diamond face card than a diamond, the number 0.3 must be incorrect. Since it is less likely to draw an even numbered diamond than a diamond, the number 0.15 might be correct; a calculation is needed if you want to be sure.
- Estimate the probability of getting a red face card. Solution: The probability of a red card is 0.5, so the probability of a red face card must be less than 0.5. We can do even better, because we know that the face cards are less than half the cards in a suit. So the probability of a red face card must be less than 0.25.

Bonferroni's Inequality

Another relationship of probabilities is Bonferroni's inequality, which relates the probability of the events A, B, and $A \cap B$.

> **NOTE**
>
> **Bonferroni's Inequality**
>
> If A and B are events in the same sample space, then $P(A \cap B) \geq P(A) + P(B) - 1$.

This inequality follows from the General Addition Rule. For two events A and B, we have $P(A \cup B) = P(A) + P(B) - P(A \cap B)$.

Add $P(A \cap B)$ to both sides to get $P(A \cap B) + P(A \cup B) = P(A) + P(B)$.

Subtract $P(A \cup B)$ from both sides, now getting $P(A \cap B) = P(A) + P(B) - P(A \cup B)$.

We don't know much about $P(A \cup B)$ except that, like any other probability, it can't be greater than 1, so $P(A \cup B) \leq 1$. Multiply both sides of this inequality by -1, and you get $-1 \leq -P(A \cup B)$; remember to reverse the sense of an inequality when multiplying by a negative number. Substituting this in the equation for $P(A \cap B)$ given previously leads to $P(A \cap B) = P(A) + P(B) - P(A \cup B) \geq P(A) + P(B) - 1$.

Drop the middle term, and you have Bonferroni's Inequality, named after the Italian mathematician Carlo Emilio Bonferroni, who discovered it in 1935. This inequality is most useful when $P(A)$ and $P(B)$ are close to 1. It can be extended to more than two events.

Examples of Bonferroni's Inequality

- Draw one card from a standard deck. The probability of getting a card that is not an ace is $48 / 52 \approx 0.923$, and the probability of getting a card that is not a club is $39 / 52 = 0.75$. The probability of getting a card that is not an ace and not a club is at least $0.923 + 0.75 - 1 = 0.673$. In fact, since there are 39 cards that are not clubs and three of these are aces, it is $(39 - 3) / 52 \approx 0.692$.
- Toss four dice. By the Complement Rule, the probability of getting at least one even number is $1 - (3 / 6)^4 \approx 0.938$, and the probability of getting at least one pair is $1 - (6 \times 5 \times 4 \times 3 / 6^{4}) \approx 0.722$. The probability of getting at least one even number and at least one pair is at least $0.938 + 0.722 - 1 = 0.660$.

5.3 Problems Using Dice

A die has six sides or faces, numbered 1, 2, 3, 4, 5, and 6. Assume that each face is equally likely to appear when throwing a die. In many games, two dice are thrown together. When two dice are thrown together, we can write (1, 2) for the event of 1 on the first die and 2 on the second die. This is a different event from (2, 1), the event of 2 on the first die and 1 on the second die.

Examples of Dice Problems

- One die is thrown. What is the probability of getting a 4? Solution: There are six equally likely outcomes, and there is one success. Thus $1 / 6 \approx 0.167$ is the probability.
- One die is thrown. What is the probability of getting an even number? Solution: There are three successes, so $3 / 6 = 0.5$ is the probability.
- Two dice are thrown. What is the probability of getting a sum of 5? Solution: There are $6 \times 6 = 36$ possible outcomes. Of these, there are four ways to get a sum of 5: (1, 4), (2, 3), (3, 2), (4, 1). Thus $4 / 36 \approx 0.111$ is the answer.
- Two dice are thrown. What is the probability of getting at least one 5? Solution: There are three outcomes we must count: 5 on the first die and something different on the second die, 5 on the second die and something different on the first, and 5 on both dice. Thus there are $5 + 5 + 1 = 11$ outcomes in the event "at least one 5," and $11 / 36 \approx 0.306$ is the answer. You can also solve this by finding the complement of the event that no 5s appear. There are $5 \times 5 = 25$ ways to get both numbers different from 5, so the probability of at least one 5 is $1 - (25 / 36) \approx 0.306$.
- Two dice are thrown. What is the probability of getting at least one odd number? Solution: This is another example of a situation where it is easier to look at the *complement* of the desired event than the event itself. The complement of this is the event that both dice are even. There are 3×3 ways to get even numbers on both dice, so the probability of both being even is $9 / 36 = 0.25$, and we get $1 - 0.25 = 0.75$ for the probability of at least one odd number appearing.
- Two dice are thrown. Let A be the event that the sum of the numbers on the two dice is even and let B be the event that at least one number 6 appears. Describe and give the probabilities of the events A, B, $A \cap B$, $A \cup B$, $A \cap B^C$, and $A \setminus B$.

  Solution: The event A that the sum is even can happen in 18 ways. For each number on the first die, there are three numbers on the second die that give an even sum. So $P(A) = 0.5$.

- The event B is the event that at least one 6 appears. The reasoning is the same as in a previous example, where the event that at least one 5 appears, and P(B) is about 0.306.
- The event $A \cap B$ is the event that the sum is even and at least one 6 appears. This is the same as 6 on one die and an even number on the other. There are five ways this can happen: (2, 6), (4, 6), (6, 2), (6, 4), and (6, 6), so $5 / 36 \approx 0.139$ is the probability.
- The event $A \cup B$ is the event that the sum is even or one 6 appears. There are 24 ways this can happen: (1, 1), (1, 3), (1, 5), (1, 6), (2, 2), (2, 4), (2, 6), (3, 1), (3, 3), (3, 5), (3, 6), (4, 2), (4, 4), (4, 6), (5, 1), (5, 3), (5, 5), (5, 6), (6, 1), (6, 2), (6, 3), (6, 4), (6, 5), and (6, 6). Thus, $24 / 36 \approx 0.667$ is the probability of $A \cup B$. We could also solve this using the General Addition Rule, getting $P(A) + P(B) - P(A \cap B) = 0.5 + 0.306 - 0.139 \approx 0.667$.
- The event $A \cap B^C$ is that the sum is even and that 6 does not appear. There are 13 ways that this can happen: (1, 1), (1, 3), (1, 5), (2, 2), (2, 4), (3, 1), (3, 3), (3, 5), (4, 2), (4, 4), (5, 1), (5, 3), and (5, 5). This gives $13 / 36 \approx 0.361$ for the probability of $A \cap B^C$.
- The event $A \setminus B$ is the same as the event $A \cap B^C$ as can be seen by drawing a Venn diagram and its probability is the same, approximately 0.361.

5.4 Problems Involving Coin Tossing

When a fair coin is tossed, you assume that you have equal chances of getting a head or getting a tail. That means P(H) = 0.5 and P(T) = 0.5 for each toss. Most coin tossing problems involve tossing more than one coin or tossing a coin many times in succession.

Examples of Coin Tossing Problems

- What is the probability of getting at least one head when tossing three coins? Solution: Write out the sample space, and then see how many outcomes are in the event of at least one head. The sample space is {HHH, HHT, HTH, HTT, THH, THT, TTH, TTT}, and the desired event includes every outcome except the last, so 7 / 8 = 0.875 is the probability.
- What is the probability of getting the sequence HTTHT when you toss a coin five times? Solution: The probability of H on the first toss is 1 / 2 = 0.5, T on the second toss also has probability 0.5, and so on. The probability is $(0.5)^5 = 0.03125$ for the given sequence.
- What is the probability of getting exactly two heads when two coins are tossed? Solution: The sample space is {HH, HT, TH, TT}, and each outcome is equally likely. For HH, 1 / 4 or 0.25 is the probability.
- A coin is tossed six times in succession. What is the probability of getting exactly two heads and four tails, in any order? Solution: There are $2^6 = 64$ different elements in the sample space. There are ${}_6C_2$ ways to choose the places where the heads will occur. All the other places must be tails. Since ${}_6C_2 = 15$, 15 / 64 = 0.234375 is the probability of getting two heads and four tails.
- A coin is tossed until a head appears. What is the probability that this will take four throws? At most four throws? Solution: The sample space for this experiment has infinitely many outcomes in it, so we need to use the Multiplication Principle to find the probability of each possible outcome. The outcomes that are relevant are H, TH, TTH, and TTTH. The probability of H on the first throw is 1 / 2; of getting TH on the first two throws is (1 / 2) × (1 / 2); for TTH on the first three throws, it is (1 / 2) × (1 / 2) × (1 / 2); and for TTTH on the first four throws, it is (1 / 2) × (1 / 2) × (1 / 2) × (1 / 2). For getting the first head on the fourth throw, the probability is 1 / 16, or 0.0625. There are four mutually exclusive events here (getting the first head on the first, second, third, or fourth throw), so we use the Addition Rule for Mutually

Exclusive Events. Thus getting the first head in at most four throws has probability equal to the sum of the probabilities for getting a head on the first, second, third, or fourth throw, so we get

$$\frac{1}{2}+\frac{1}{4}+\frac{1}{8}+\frac{1}{16}=\frac{15}{16}=0.9375$$

for the probability of a head in at most four tosses.

5.5 Problems Involving Cards

Card problems are a good source of probability problems because there are so many different variations and possibilities. A standard deck has 52 cards, with four suits: hearts, diamonds, clubs, and spades. Cards in each suit are numbered from 1 to 10 with three face cards, a king, queen, and jack in each suit. The card numbered 1 is called the ace.

Cards are dealt at random for the games of poker, where a hand has five cards, and bridge, where four players each have a hand of 13 cards.

Useful numbers to know: There are ${}_{52}C_5 = 2{,}598{,}960$ different possible poker hands and ${}_{52}C_{13} = 635{,}013{,}559{,}600$ different possible bridge hands.

Examples of Card Problems

- One card is drawn from a deck of 52. What is the probability that it is a face card? Solution: There are four suits with three face cards in each, so $12 / 52 \approx 0.231$ is the probability of drawing a face card.
- One card is drawn from a deck of 52. What is the probability that it is a red card? Solution: There are two red suits with 13 cards each, so $26 / 52 = 0.5$ is the probability of drawing a red card.
- One card is drawn from a deck of 52. What is the probability that it is a red 10? Solution: There are two red tens, 10♥ and 10♦, so $2 / 52 \approx 0.0385$ is the probability of drawing a red 10.

- What is the probability of getting a royal flush—ace, king, queen, jack, and 10 of one suit—in poker? Solution: There are four possible royal flushes, one for each suit, so the probability of getting a royal flush is $4 / 2{,}598{,}960 \approx 1.539 \times 10^{-6}$, which is less than two in a million.
- What is the probability of getting exactly two aces in a bridge hand? Solution: There are ${}_4C_2 = 6$ ways of choosing the two aces. Once the aces have been chosen, there are 48 cards left to choose from because we can't choose one of the other aces. We have ${}_{48}C_{11} = 22{,}595{,}200{,}368$ ways to choose the other 11 cards in the hand. There are then $6 \times 22{,}595{,}200{,}368$ possible hands with exactly two aces and we get $6 \times 22{,}595{,}200{,}368 / 635{,}013{,}559{,}600 \approx 0.213$ for the probability of getting exactly two aces.

5.6 Geometric Probability

In geometric probability, the sample space is a region, like a line, curve, area, or volume. An event is a region inside the sample space; this region is called the **feasible region**. The probability of choosing a point at random in the feasible region is the ratio of the length (or area or volume) of the feasible region divided by the length (or area or volume) of the whole sample space.

For example, a point is chosen at random on an 8" × 10" computer screen that has half the area on the screen black and half white. The probability of choosing a white point is 0.5 and is the same for a black point.

Examples of Geometric Probability

- A mark is made on an 11-inch strip of paper at random. What is the probability that the mark is within 2 inches of either end? Solution: There is a 4-inch region on the 11-inch strip that is within 2 inches of an end, so the feasible region has length 4 inches. The probability of the random mark being in the feasible region is $4 / 11 \approx 0.364$.

- A dart is thrown at a target of radius 10. The bull's-eye has radius 2. If you hit the target at random, what is the chance of hitting the bull's-eye? Solution: The area of the target is $\pi \times 10^2$ and the area of the bull's-eye is $\pi \times 2^2$, so the probability of hitting the bull's-eye is $(\pi \times 2^2) / (\pi \times 10^2) \approx 0.04$.

Examples of Time Interval Problems

The method of geometric probability can also be applied to time intervals.

- Kane's father picks him up from school at random between 3:00 and 3:30. What is the probability that he will pick up Kane between 3:10 and 3:20 today? Solution: There is a 30-minute time period for the sample space and a 10-minute interval for the feasible region. The probability is $10 / 30 \approx 0.333$ that Kane's father will come between 3:10 and 3:20.
- A classroom computer is available for student use between 11:00 and 12:00. Cara needs to work on the computer for 10 minutes and will choose the time she goes to the computer at random. Easton plans to start using the computer sometime between 11:00 and 12:00. What is the probability that the computer will be free when he arrives? Solution: Once Cara has chosen her 10 minutes, Easton has a 50-minute interval available for his arrival, so this is the feasible region. Thus the probability is $50 / 60 \approx 0.833$ that Easton will find the computer free.

5.7 Quality Control

Some manufactured products turn out to be defective. For example, a gear might be too small or a light bulb might not last as long as it should. Quality control is a process that businesses use to reduce errors that cause defective products.

The causes of defective products might be variations in the quality of the raw materials, defective performance in the manufacturing equipment, human error, and so on. But whatever the cause, it often seems like defective products appear at random. This means that probability is a useful tool in quality control.

One feature of quality control is the testing of individual products as they come off the assembly line. Sometimes, every product is tested. But in other cases, testing requires the destruction of the product or may otherwise be costly. For example, if a light bulb is tested to see how long it lasts or if a beam is stressed until it breaks, the product is destroyed. Measuring the size of ball bearings to make sure they meet a buyer's requirements may have a high cost in comparison to the cost of producing the ball bearings. In such cases, testing is done on a random sample rather than on every item produced.

Examples of Quality Control Problems

- There are five defective products in a bin containing 100 products. Choose one product at random. What is the chance that it is defective? Solution: There are 100 elements in the sample space of all products and five outcomes that are defective products, so the probability of getting a defective product is 5 / 100 = 0.05.
- Now look at the same situation, where there are five defective products in a bin of 100, but choose two products at random. What is the chance that one is defective? That both are defective? That at least one is defective? This is called **sampling without replacement**, which means that the second product is chosen without putting the first back in. Solution: There are ${}_{100}C_2 = 4{,}950$ ways to choose two items, so this is the size of the sample space. There are $95 \times 5 = 475$ ways of getting one defective and one that is not defective, so the probability of one defective is 475 / 4,950 $\approx$ 0.0960. There are ${}_5C_2 = 10$ ways of getting two defective products with a probability of 10 / 4,950 $\approx$ 0.002. The probability of getting at least one defective is the sum of the probabilities for these two mutually exclusive events and is $0.096 + 0.002 \approx 0.098$.

- Again, with 100 items of which five are defective, suppose one item is chosen at random, tested, and put back in, and then another is chosen at random, tested, and put back in. What is the probability that one is defective? Both? At least one? This is called **sampling with replacement**, which means that the second product is chosen after the first one is put back in. Solution: If only one item is chosen, the probability that it is defective is ${}_5C_1 / 100 = 5 / 100 = 0.05$, and the probability that it is not defective is ${}_{95}C_1 / 100 = 0.95$. The probability that one is defective when two are sampled with replacement is the sum of two mutually exclusive events, the first defective and the second not defective and the first not defective and the second defective. So the probability of exactly one defective is $(0.05 \times 0.95) + (0.95 \times 0.05) = 0.095$. The probability that both are defective is $0.05 \times 0.05 = 0.0025$. The probability that at least one is defective is the sum of these two probabilities, $0.095 + 0.0025 = 0.0975$.

Notice that the answers for sampling with replacement and sampling without replacement are close to each other. You can see that the computation for sampling with replacement is easier. However, the two methods have different statistical properties, and sometimes one method is preferred to the other in statistics.

5.8 Birthday Problems

There are many different problems based on birthdays—whether people share the same birthday or birth month, or whether they have different birthdays. In such problems, you usually assume that all days of the year are equally likely as birthdays, and that all months have the same length and ignore the day added in leap years. The numbers are large and a calculator or computer is necessary.

Examples of Birthday Problems

- In a group of 23 people, what is the probability that at least two people have the same birthday? Solution: It is easier to look at the *complement* of the event than the event itself. That means we start by looking for the probability that everyone has a different birthday. We reason using an ordering of the people. There are 365^{23} ways of choosing a birthday for each individual (this number tells how many elements are in the sample space), and there are ${}_{365}P_{23}$ ways of choosing *different* birthdays for everyone (this counts the outcomes in the event). Thus the probability that each person has a different birthday is

$$\frac{{}_{365}P_{23}}{365^{23}} \approx \frac{4.220088 \times 10^{58}}{8.5651679 \times 10^{58}} \approx 0.492705$$

- The probability that at least two people have the same birthday is then $1 - 0.4927025 \approx 0.5072975$, or a little better than half.
- In a group of three friends, what is the probability that they were born on different days of the week? Solution: There are $7^3 = 343$ elements in the sample space; each element is a possible way that the friends can have their birthdays during the week. There are $7 \times 6 \times 5 = 210$ ways that the birthdays can be different, so the probability is $210 / 343 \approx 0.612$ that the three friends were born on different days of the week.

5.9 Transmission Errors

Codes are used to transmit data. For example, the letters in an e-mail message are converted into a string of 1s and 0s before sending.

During transmission, errors occur at random. For a specific communication channel, the error rate is determined empirically, by looking at actual transmissions and dividing the number of errors by the total number of bits (letters or numbers) sent. You can then use this error rate to predict what might happen in another transmission.

Examples of Transmission Errors

- The error rate for a certain type of transmission is 0.01 (which is the probability of an error). What is the probability that there will be two errors in a message of 30 bits? Solution: There are ${}_{30}C_2 = (30 \times 29) / 2! = 435$ ways to choose two places for the errors. Each of these two places has probability 0.01 for an error. Each of the other 28 places has 0.99 probability of no error. Therefore, the probability of exactly two errors is $(.01)^2 \times 435 \times (.99)^{28} \approx 0.0328$.
- The error rate for another type of transmission is 0.20, which is quite high compared to the previous example. For a better chance of sending an accurate message over such a noisy channel, you can send your message multiple times. To send the message 0, you might transmit the code word 000, and to send the message 1, you might transmit the code word 111. Suppose the receiver gets the code word 010. What is the probability that the message 0 was received as the code word 010? What is the probability that the message 1 was received as 010? What message does the recipient think was sent? Solution: If the message 0 was sent, the code word was 000. The probability that this was changed to 010 is $0.8 \times 0.2 \times 0.8 = 0.128$. The probability that the code word 111 was changed to 010 is $0.2 \times 0.8 \times 0.2 = 0.032$. It is four times more likely that the message 0 and code word 000 was transmitted to the recipient than the message 1 and code word 111.

Chapter 5 Summary

A **sample space** is the set of all possible outcomes of an experiment or observation. An **event** is a specific subset of the sample space. A **trial** is one performance of the experiment or observation.

If all outcomes in the sample space are equally likely, the probability of an event is

$$\frac{\text{number of elements in the event}}{\text{number of elements in the sample space}}$$

The **significant digits** of a number are all the digits in the number except zeros at the beginning of the number or at the end of the number.

Rounding gives a number with fewer significant digits. To minimize the loss of accuracy, round only at the conclusion of a series of computations.

A number in **scientific notation** is written as a product of a number between 0 and 10 (or between 0 and -10 if the number is negative) times a power of 10.

NOTE

Multiplication Rule for Probability

If we perform two trials and the probability of a specified outcome in the first trial is p and the probability of another specified outcome in the second trial is q, then the probability of both outcomes happening in two trials is pq.

Tree diagrams give a picture of all possible outcomes of an experiment or observation.

NOTE

Complement Rule

For an event A, $P(A^C) = 1 - P(A)$.

NOTE

Addition Rule of Probability for Mutually Exclusive Events

If A and B are mutually exclusive events in the same sample space, then $P(A \cup B) = P(A) + P(B)$.

NOTE

General Addition Rule

If A and B are events in the same sample space, then $P(A \cup B) = P(A) + P(B) - P(A \cap B)$.

NOTE

Probability Inequality

If event A is a subset of event B, then $P(A) \leq P(B)$.

NOTE

Bonferroni's Inequality

If A and B are events in the same sample space, then $P(A \cap B) \geq P(A) + P(B) - 1$.

A **die** has six sides or faces, numbered 1, 2, 3, 4, 5, and 6. It is assumed that each face is equally likely to appear when throwing a die. Usually, several dice are thrown together.

When a fair **coin** is tossed, assume that $P(H) = 0.5$ and $P(T) = 0.5$ for each toss.

A standard deck of **cards** has 52 cards, with four suits of hearts, diamonds, clubs, and spades. Each suit has cards numbered from 1 to 10 and three face cards, king, queen, and jack. The card numbered 1 is called the ace.

A **poker hand** has five cards and a **bridge hand** has 13 cards. There are ${}_{52}C_5 = 2{,}598{,}960$ different poker hands and ${}_{52}C_{13} = 635{,}013{,}559{,}600$ possible bridge hands.

In **geometric probability**, the sample space is a region, and an event, called the **feasible region,** is a region inside the sample space. The probability of choosing a point at random in the feasible region is the size of the region divided by the size of the sample space.

The **error rate** in a transmission channel is determined empirically as number of errors / number of bits transmitted.

Chapter 5 Practice Problems

1. How many significant digits does each of the following have?
   a. 5.3407
   b. 5,670,000
   c. 3,000.4
   d. 6.80
   e. 100
   f. 0.00053
2. Round each of the following to three decimal places.
   a. 31.78865
   b. 0.00004
   c. 6.32
   d. 4.55555
3. Write each of the following in scientific notation. Round to three significant digits.
   a. 72,338,525
   b. 0.0007658
   c. 5,274,364,211.005
4. For each of the following, convert from scientific notation to an ordinary number.
   a. 3.056×10^6
   b. 1.3×10^{-3}
   c. 4.242×10^4
   d. 5.002×10^{-7}
5. A number is chosen at random from the primes less than 30. What is the probability that it contains the digit 2?
6. A month is chosen at random. What is the probability that the name of the month ends in the letter *y*?
7. A day of the week is chosen at random. What is the probability that the name of the day contains the letter *e*?

8. A coin is flipped and then a die is thrown. What is the probability of heads followed by an even number?
9. You choose a letter at random from the word *green* and then a letter from the word *glasses*. What is the probability of getting the letter *e* followed by the letter *s*? Draw a tree diagram to illustrate your answer.
10. You choose a month at random and then a day of the week at random. What is the probability that you will get a Saturday in May? Illustrate your answer with a tree.

Dice Problems

11. One die is thrown. What is the probability of getting a 3?
12. One die is thrown. What is the probability of getting an odd number?
13. Two dice are thrown. What is the probability of getting a sum that is equal to 6?
14. Two dice are thrown. What is the probability of getting doubles?
15. Two dice are thrown. What is the probability of getting a sum of at least 5?
16. Three dice are thrown. What is the probability of getting at least one pair?
17. Two dice are thrown. What is the probability of getting doubles or two numbers that differ by 4?
18. Two dice are thrown. What is the probability of getting two odd numbers or two numbers that differ by 4?
19. When throwing two dice, which is greater, the probability of getting doubles or getting an even sum?
20. Give a lower estimate for the probability of at least one even and at least one greater than 2 when throwing two dice.

Coin Problems

21. What is the probability of getting at least one tail when tossing four coins?
22. What is the probability of getting the sequence HTTT when a coin is tossed four times?

23. What is the probability of getting exactly three heads when five coins are tossed?

24. A coin is tossed until two heads appear. What is the probability that this will take four throws? At most four throws?

Card Problems

25. One card is drawn from a standard deck of 52. What is the probability that it is a 10?

26. One card is drawn from a deck. What is the probability that it is a black king?

27. One card is drawn from a deck. What is the probability that it is not a club?

28. Three cards are drawn from a deck. What is the probability that none of them is a face card?

29. Four cards are drawn from a deck. What is the probability of getting exactly two hearts?

30. One card is drawn from a deck. What is the probability of getting a face card or a black ace?

31. What is the probability of getting a four of a kind—one from each suit with the same value and any other card—in poker?

32. Two cards are drawn from a deck. Which is greater, the probability of getting a pair (two cards with the same value) or two kings?

Quality Control Problems

33. Three items are chosen from a bin containing 90 items, of which six are defective and then tested. (This is sampling without replacement.)

   a. What is the probability that one is defective?

   b. What is the probability that two are defective?

   c. What is the probability that three are defective?

   d. What is the probability that at least one is defective?

34. Three items are chosen and tested one by one from a bin containing 90 items, of which six are defective; after each item is chosen, it is replaced. (This is sampling with replacement.)

a. What is the probability that one is defective?

b. What is the probability that two are defective?

c. What is the probability that three are defective?

d. What is the probability that at least one is defective?

Birthday Problems

35. What is the probability that in a group of 30 people at least two people share the same birthday?

36. What is the probability that in a group of five people at least two people have birthdays in the same month? (Assume that each month is equally likely to be a birthday for each person, even though the months have different lengths.)

37. What is the probability that in a group of 12 people, every birthday is in a different month? (Assume that each month is equally likely to be a birthday for each person, even though the months have different lengths.)

38. What is the probability that in a family of five, all birthdays will be in just two different months? (Assume that each month is equally likely to be a birthday for each person, even though the months have different lengths.)

Geometric Probability Problems

39. A point is chosen at random in a square. What is the probability that it is contained in a circle inscribed in the square?

40. The midpoints of the sides of an equilateral triangle are connected to one another, forming four new smaller triangles, each similar to the original triangle. A point is chosen at random. What is the probability that it will be in the small triangle in the center?

41. Ryker likes to watch the news each evening from 8:00 to 8:30 p.m., and his friend Grady likes to call him at random sometime between 6:30 and 10:00 p.m. What is the probability that Grady will call when Ryker is watching the news?

Transmission Error Problems

42. The error rate of a type of transmission is 0.15 errors per bit.

a. What is the probability of exactly two errors in a message of 10 bits?

b. What is the probability of exactly eight errors in a message of 10 bits?

43. The error rate for one type of transmission is 0.30. To send the message 0, you transmit the code word 00000, and to send the message 1, you transmit the code word 11111. Suppose the receiver gets the code word 10011. What is the probability that the message 0 was intended? What is the probability that the message 1 was intended? What message do you think was intended?

More Challenging Problems

44. Five pairs of twins attend a special event for twins. Two of them are chosen at random to win the door prize. What is the probability that the two are siblings?

45. Silas has 12 pairs of shoes in his closet. One morning, he chooses four shoes at random. What is the probability that there is at least one matching pair of shoes in the four he has chosen?

46. Two dice are thrown and then a card is selected at random from a deck of 52. What is the probability that the sum of the numbers on the dice will be the same as the number on the card? For this problem, count the ace as 1.

47. There are four keys and four locked boxes; each of the keys opens one box. In one of the boxes is $1,000, and in each of the other boxes is a lump of coal. You can choose one of the keys and one of the boxes. What is the chance that the key will open the box? What is the chance that you will win the $1,000?

48. Suppose you can choose between two games. In one game, you win $10 if you throw six dice and get at least one 6; in the other game, you win $10 if you throw 12 dice and get at least two 6s. Which is the better game?

49. Is it more likely to get at least one 6 when throwing four dice or to get at least one pair of 6s in 24 throws of two dice? (This is known as de Méré's paradox for the French writer Chevalier de Méré who brought the problem to the attention of mathematicians.)

50. Of the last 12 calls that Mercedes has received from her cousin, all were on Mondays or Wednesdays. Is Mercedes correct in thinking the calls are not at random?
51. Of the last 12 calls that Mercedes received from her aunt, none was on a Sunday. Is Mercedes correct in thinking the calls are not at random?
52. A bag of 100 candies contains exactly 10 mints. You choose 10 at random. What is the probability that you get no mints?
53. You are given seven keys, and you know that one of them opens a treasure chest. You try one key after another, in random order. What is the chance that you will open the chest on the first try? Second try? Third try? What pattern do you see? Can you justify the pattern that you see?

Answers to Chapter 5 Practice Problems

1. **a.** 5
   b. 3
   c. 5
   d. 3
   e. 3
   f. 2
2. **a.** 31.789
   b. 0.000
   c. 6.32. You cannot increase the number of significant digits in the rounding process. So the number you started with, 6.32, cannot be rounded to three decimal places.
   d. 4.556
3. **a.** 7.23×10^7
   b. 7.66×10^{-4}
   c. 5.27×10^9

4. **a.** 3,056,000
   b. 0.0013
   c. 42,420
   d. 0.0000005002
5. The primes less than 30 are 2, 3, 5, 7, 11, 13, 17, 19, 23, and 29. The probability that the digit 2 appears in the number is 3 / 10 = 0.3.
6. There are four months that end in the letter *y*: January, February, May, and July. The probability is 4 / 12 ≈ 0.333.
7. There are two days that contain the letter *e*: Tuesday and Wednesday. The probability is 2 / 7 ≈ 0.286.
8. The probability of heads is 0.5 and the probability of an even number is 0.5; so the probability that both will happen is 0.5 × 0.5 = 0.25.
9. The probability of letter *e* from the word *green* is 2 / 5 = 0.4 and the probability that it is followed by letter *s* from *glasses* is 3 / 7 ≈ 0.429, as shown in the tree in Figure 5.6. Thus the probability of getting an *e* followed by an *s* is 0.4 × 0.429 ≈ 0.172. The four probabilities in the figure do not add up to 1 because of a rounding error. The tree has been simplified by grouping together all the letters that are not in the events we are interested in.

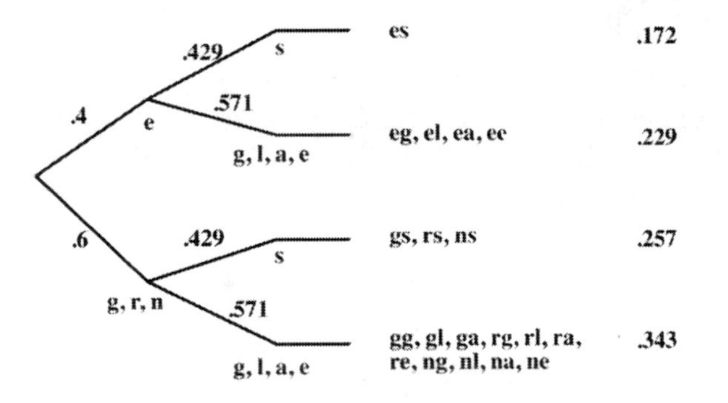

Figure 5.6
Probability tree for problem 9.

10. The probability of getting May is $1 / 12 \approx 0.083$ and the probability of Saturday is $1 / 7 \approx 0.143$, so the probability of a Saturday in May is 0.012. This is shown in the tree in Figure 5.7, where the months other than May have been grouped together, and the days other than Saturday have been grouped together.

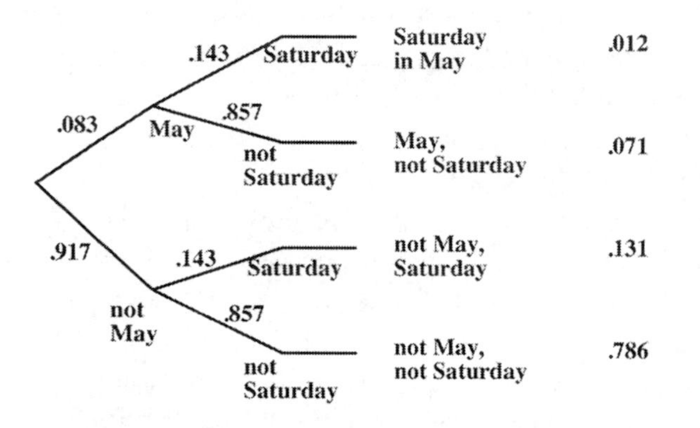

Figure 5.7
Probability tree for problem 10.

11. There are six equally likely outcomes, and there is one success. Thus $1 / 6 \approx 0.167$ is the probability.

12. There are three successes, so $3 / 6 = 0.5$ is the probability.

13. There are 36 possible outcomes. Of these, there are five ways to get a sum of 6: (1, 5), (2, 4), (3, 3), (4, 2), and (5, 1), with the first number on the first die and the second number on the second die. Thus $5 / 36 \approx 0.139$ is the answer.

14. There are 36 possible outcomes. Of these, there are six ways to get doubles. Thus $6 / 36 \approx 0.167$ is the answer.

15. It is easier to look at the complement and to count up the number of pairs that add up to less than 5: (1, 1), (1, 2), (1, 3), (2, 1), (2, 2), (3, 1). There are six such pairs, so there are 30 pairs adding up to 5 or more, and $30 / 36 \approx 0.833$ is the probability for getting a sum that is at least 5.

16. It is easier to use the complement. The probability of getting all different dice is $(6 \times 5 \times 4) / 6^3 \approx 0.556$, so the probability of at least one pair is about 0.444.
17. The probability of getting doubles is $6 / 36 \approx 0.167$. The pairs that differ by 4 are (1, 5), (2, 6), (5, 1), and (6, 2); the probability of such a pair is $4 / 36 \approx 0.111$. These two events are disjoint, so the final answer is approximately 0.278.
18. These events are not disjoint. The probability of getting both odd numbers is $(3 \times 3) / 36 = 0.250$, the probability of two numbers that differ by 4 is approximately 0.111 from the previous problem. The probability of two odd numbers that differ by 4 is $2 / 36 \approx 0.056$, since there are only two such pairs, (1, 5) and (5, 1). The answer is approximately $0.250 + 0.111 - 0.056 \approx 0.305$.
19. Since doubles always give an even sum, the set of outcomes that are doubles is a subset of the set of outcomes that give an even sum. Therefore, the probability of getting an even sum is greater.
20. Use Bonferroni's Inequality. To get the probability of getting at least one even, use the Complement Rule, $1 - (3 \times 3) / 36 = 1 - 0.25 = 0.75$. Also use the Complement Rule to find the probability of at least one die greater than 2, $1 - (2 \times 2 / 36) \approx 1 - 0.111 \approx 0.889$. The probability of getting at least one even at least one greater than 2 must be bigger than $0.75 + 0.889 - 1 \approx 0.639$.
21. The sample space is {HHHH, HHHT, HHTH, HTHH, THHH, HHTT, HTHT, HTTH, THHT, THTH, TTHH, HTTT, THTT, TTHT, TTTH, TTTT}. The probability of getting all heads is $1 / 16 = 0.0625$ so $1 - 0.0625 = 0.9375$ is the probability of at least one tail.
22. $(1 / 2)^4 = 0.0625$
23. ${}_5C_3 / 2^5 = 10 / 32 = 0.3125$
24. Successful outcomes with their probabilities are HH, $1 /4 = 0.25$; HTH, $1 / 8 = 0.125$; THH, $1 / 8 = 0.125$; HTTH, $1 / 16 = 0.0625$; THTH, $1 / 16 = 0.0625$; and TTHH, $1 / 16 = 0.0625$. The probability of getting two heads only after four throws is $0.0625 + 0.0625 + 0.0625 = 0.1875$. The probability of getting two heads in at most four throws is $0.25 + 0.125 + 0.125 + 0.0625 + 0.0625 + 0.0625 = 0.6875$.

25. $4 / 52 \approx 0.077$

26. $2 / 52 \approx 0.038$

27. $39 / 52 = 0.75$

28. There are 12 face cards and 40 cards that are not face cards. There are ${}_{40}C_3 = 9880$ ways to get three cards with none being a face card. There are ${}_{52}C_3 = 22{,}100$ elements in the sample space. The answer is $= 9{,}880 / 22{,}100 \approx 0.447$.

29. There are ${}_{13}C_2$ ways to get two hearts, and ${}_{39}C_2$ ways to get two cards that are not hearts. There are ${}_{52}C_4$ ways to choose four cards, so $({}_{13}C_2 \times {}_{39}C_2) / {}_{52}C_4 \approx 0.213$ is the probability of exactly two hearts.

30. The probability of a face card is $12 / 52 \approx 0.231$ and the probability of a black ace is $2 / 52 \approx 0.038$. Since these are disjoint events, we can add the probabilities, getting about 0.269 for the final answer.

31. There are 13 ways to get four of a kind; after that, there are 48 ways to choose the fifth card, so there are $13 \times 48 = 624$ different hands with four of a kind. Dividing by the number of possible poker hands, we get $624 / 2{,}598{,}960 \approx 2.40 \times 10 - 4$ as the probability of a flush, about two in 10,000.

32. Use the Inequality Property. The event of two kings is a subset of the event of a pair. Thus, the probability of drawing two kings is less than drawing a pair.

33. There are ${}_{90}C_3 = 117{,}480$ elements in the sample space.

a. There are ${}_{6}C_1 = 6$ ways to get one defective and ${}_{84}C_2 = 3{,}486$ ways to get two non-defectives, so the chance of getting exactly one defective is $(6 \times 3{,}486) / 117{,}480 \approx 0.178$.

b. There are ${}_{6}C_2 = 15$ ways to get two defectives and ${}_{84}C_1 = 84$ ways to get one non-defective, so the chance of getting exactly two defectives is $(15 \times 84) / 117{,}480 \approx 0.011$.

c. There are ${}_{6}C_3 = 20$ ways to get three defectives, so the chance of getting all three defectives is $20 / 117{,}480 \approx 0.00017$.

d. The probability of at least one defective is $0.178 + 0.011 + 0.00017 \approx 0.189$.

34. The probability of a defective is 0.0667 and the probability of a non-defective is 0.9333.

a. ${}_3C_1 \times 0.0667 \times 0.9333 \times 0.9333 \approx 0.174$

b. ${}_3C_2 \times 0.0667 \times 0.0667 \times 0.9333 \approx 0.0125$

c. The probability of all three defective is $6^3 / 729{,}000 \approx 0.0667^3 \approx$ 0.000296.

d. The probability of at least one defective is $0.174 + 0.0125 + 0.000269 \approx 0.187$.

35. There are $365^{30} \approx 7.392 \times 10^{76}$ elements in the sample space. There are ${}_{365}P_{30} \approx 2.171 \times 10^{76}$ ways the 30 birthdays could be different. The probability that each person has a different birthday is $2.171 \times 10^{76} / 7.392 \times 10^{76} \approx 0.294$, so the probability that at least two people share a birthday is $1 - 0.294 \approx 0.706$, which is quite likely.

36. There are ${}_{12}P_5 = 95{,}040$ ways to choose different months for five people and $12^5 = 248{,}832$ ways to choose months for the birthdays for five people. The probability of all five people having birthdays in different months is $95{,}040 / 248{,}832 \approx 0.382$. Subtracting this from 1 gives the probability that the five people are *not* in different months: $1 - 0.382 \approx 0.618$.

37. There are ${}_{12}P_{12} \approx 4.790 \times 10^8$ ways to choose months for 12 people and $12^{12} \approx 8.916 \times 10^{12}$ ways to choose months for the birthdays for 12 people. The probability of all 12 people having different birthdays is $(4.790 \times 10^{8}) / (8.916 \times 10^{12}) \approx 5.37 \times 10^{-5}$.

38. There are $12^5 = 248{,}832$ ways to choose months for five people. There are ${}_{12}C_2 = 66$ ways to choose the two months. There are $2^5 = 32$ ways to put the five birthdays into the two months. However, two of these ways have everyone in one or the other month. So there are $32 - 2 = 30$ ways to choose the birthdays in two months. The probability that the five birthdays are in two months is $(66 \times 30) / 248{,}832 \approx 0.007957$, which is very small.

39. For this problem, it doesn't matter what size the square is. The inscribed circle will always have the same proportion of the square. For simplicity, let's assume the square has side equal to 1 and area equal to 1. The diameter of the circle is 1 so the radius is 0.5 and the area is $\pi\ (0.5)^2 \approx 0.785$. The probability of choosing a point in the square is $0.785 / 1 \approx 0.785$.

40. The small triangle in the center has one quarter of the area of the large triangle, so the probability of choosing a point there is $1 / 4 = 0.25$.

41. The sample space is three and a half hours, or 210 minutes. The feasible region, the time when the news is on, is 30 minutes. The probability that Grady will call while Ryker is watching the news is $30 / 210 \approx 0.143$.

42. **a.** There are ${}_{10}C_2 = 45$ ways to choose the places of the errors. The probability of getting errors in the two chosen places is $(0.15)^2 = 0.0225$ and the probability of no errors in the other eight places is $(0.85)^8 \approx 0.2725$, so the probability of exactly two errors in ten bits is $45 \times 0.0225 \times 0.272 \approx 0.276$.

b. There are ${}_{10}C_8 = 45$ ways to choose the places of the errors. The probability of getting an error in the eight chosen places is $(0.15)^8 \approx 2.563 \times 10^{-7}$ and the probability of no errors in the other two places is $(0.85)^2 = 0.7225$, so the probability of exactly two errors in ten bits is $45 \times 2.563 \times 10^{-7} \times 0.7225 \approx 8.33 \times 10^{-6}$.

43. The probability that 00000 became 10011 is $0.3 \times 0.7 \times 0.7 \times 0.3 \times 0.3 = 0.01323$ and the probability that 11111 became 10011 is $0.7 \times 0.3 \times 0.3 \times 0.7 \times 0.7 = 0.03087$. It is more than twice as likely that 1 was intended as the message.

44. There are ${}_{10}C_2 = 45$ ways to choose the two winners. Of these, five ways are siblings. So the probability of siblings winning the door prizes is $5 / 45 \approx 0.111$.

45. It is easier to use the complement of the event here. There are ${}_{24}C_4 = 10{,}626$ ways Silas can choose the shoes. There are 24 ways to choose the first shoe; then 22 ways to choose the next shoe so as not to get a pair; there are 20 ways to choose the third shoe and 18 ways to choose the fourth shoe. The order of these choices doesn't matter, so there are $(24 \times 22 \times 20 \times 18) / 4! = 7{,}920$ ways to choose the shoes *without* getting a pair. The probability of *not* getting a pair is $7{,}920 / 10{,}626 \approx 0.745$ and the probability of getting at least one pair is $1 - 0.745 \approx 0.255$.

46. There are $36 \times 52 = 1{,}872$ elements in the sample space. If the dice add up to numbers 2 through 10, the sum could match a card number. The outcomes that give 11 and 12 could not match a card and there are three such outcomes, (5, 6), (6, 5), and (6, 6). Thus, there are $36 - 3 = 33$ outcomes that could match a card. For each of these outcomes, there are four possible matches, one from each suit. Thus, there are 33×4 elements in the event. The probability of a match is thus $132 / 1872 \approx 0.071$.

47. There are $4 \times 4 = 16$ possible combinations of keys and boxes and four of them give a key that matches the lock. The chance of opening a box is $4 / 16 = 0.25$ but the chance of winning the money is $1 / 16 = 0.0625$.

48. Getting at least one 6 is the complement of getting no 6s, so the probability is $1 - (5 / 6)^6 \approx 0.665$. Getting at least two 6s is the complement of getting no 6s and getting one 6. The probability of no 6s is $(5 / 6)^{12} \approx 0.112$ and the probability of getting one 6 is $({}_{12}C_1 \times 5^{11}) / 6^{12} \approx 0.269$. The probability of at least two 6s is then approximately $1 - 0.112 - 0.269 = 0.619$. So trying for at least one 6 throwing six dice is more likely to be successful than trying for at least two 6s with 12 dice.

49. The probability of no 6s when throwing four dice is $(5 / 6)^4 \approx 0.482$, so the probability of at least one 6 is about $1 - 0.482 = 0.518$. The probability of no pairs of 6s in 24 throws is $(35 / 36)^{24} \approx 0.508$, so getting at least one pair in 24 throws is about $1 - 0.518 = 0.492$. Thus it is a little more likely to get one 6 when throwing four dice than to get two double 6s on at least one of 24 throws of two dice.

50. Compute the probability that 12 random calls would occur on a Monday or Wednesday. There are 7^{12} ways the calls could be randomly distributed over the seven days of the week. There are 2^{12} ways the calls could be randomly distributed over the two days Monday and Wednesday. The probability of all calls on Monday and Wednesday is therefore $2^{12} / 7^{12} = (2 / 7)^{12} \approx 2.959 \times 10^{-7}$, which is a very small number. Note that it is easier to compute $(2 / 7)^{12}$ than $2^{12} / 7^{12}$. Mercedes is justified in concluding that her cousin does not call at random.

51. This is similar to the previous problem. There are 6^{12} ways the calls could occur on days other than Sunday. The probability that no calls occur on Sunday is $6^{12} / 7^{12} = (6 / 7)^{12} \approx 0.157$. Mercedes is justified in concluding that her aunt does not call at random, but she is not as certain as she is about her cousin.

52. Count the elements in the event "no mints" and divide it by the number of elements in the sample space, getting ${}_{90}C_{10} / {}_{100}C_{10} \approx 0.330$.

53. The chance of opening the chest on the first try is 1 / 7. On the second try, you must multiply the probability of failure with the first key (which is 6 / 7) by the probability of success on the second key (which is 1 / 6), since you can eliminate the key you used on the first try. So the probability of success on the second try is $(6 / 7) \times (1 / 6) = 1 / 7$. Similarly, on the third try, the probability of success is $(6 / 7) \times (6 / 5) \times (1 / 5) = 1 / 7$. You can see that the probability of success is the same for every try. This makes sense because you are no more likely to find the right key on one try rather than another.

Chapter

Conditional Probability and Independence

Suppose that a dealer gives you one card from a deck of 52. You want a king. What if the dealer says that it is a club? What if the dealer tells you that it is a face card? In each case, you have more information about the card that you have been dealt. In the first case, the information doesn't help you at all. But in the second case, the information is very encouraging.

Getting a king and getting a club are **independent events**, events whose probabilities are unrelated. Getting a king and getting a face card are related events; they are *not* independent. These kinds of situations lead us to the idea of **conditional probability**, where additional information about a desired outcome changes the likelihood of getting that outcome.

In this chapter, you will learn how to compute conditional probabilities and to determine whether events are independent or not.

6.1 Conditional Probability

When you choose one card from 52, the probability of a king is $4 / 52 \approx 0.0769$ because there are four kings in a deck of 52. When you know that you have a club, the probability of a king is $1 / 13 \approx 0.0769$, since there is one king in the club suit, which has 13 cards. Both probabilities are exactly the same. Knowing that you have a club has not changed anything.

But if you know that the card you have chosen is a face card, the probability of getting dealt a king changes dramatically. In fact, it is $4 / 12 \approx 0.333$, because there are four kings among the 12 face cards. This is more than four times as likely as getting a king when you know you have a club. So in this case, the additional knowledge has completely changed the probability of getting a king.

Conditional probability handles situations where you have some additional knowledge about the outcome of a trial or experiment.

The probability of getting a king changes based on a condition—getting a face card. This condition tells us, in effect, that we have a different sample space, one that is smaller than the original sample space. The new sample space of face cards is a subspace of the old sample space of the whole deck. Our desired outcome of getting a king is an event in the new, smaller sample space; it is the intersection of the event of getting a king with the new sample space of face cards.

To understand how to get a conditional probability, look at the Venn diagram in Figure 6.1. We know that the probability of event A, P(A), is the number of outcomes in event A divided by the total number of outcomes in the sample space.

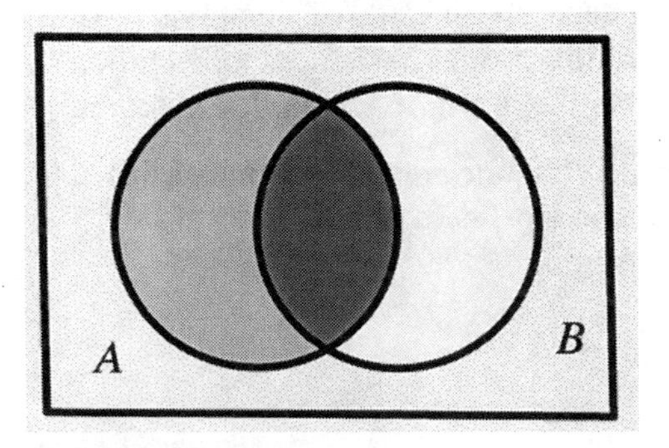

Figure 6.1
A Venn diagram showing event A and event $A \cap B$.

To compute the conditional probability of an element of A happening once we know that B has occurred, we can treat B as our new sample space. This means that we look only at the elements of the sample space that belong to event B. Instead of event A, we look at just those outcomes that are also in B, and this is the event $A \cap B$.

The probability of getting an outcome in this smaller event $A \cap B$ is the number of outcomes in $A \cap B$ divided by the number of outcomes in B. This is the same as $P(A \cap B)$ divided by $P(B)$, as shown in this equation:

$$\frac{\text{number of outcomes in } A \cap B}{\text{number of outcomes in } B} = \frac{\left(\dfrac{\text{number of outcomes in } A \cap B}{\text{number of outcomes in } S}\right)}{\left(\dfrac{\text{number of outcomes in } B}{\text{number of outcomes in } S}\right)} = \frac{P(A \cap B)}{P(B)}$$

There is a special notation for conditional probability, $P(A \mid B)$, which we read as "the probability of event A happening, knowing that event B has already happened" or "the probability of A given B." For simplicity, the intersection of two events A and B is written AB, which is read "AB" or "A and B" or "A intersect B."

With this notation, we can write the formula for conditional probability:

For two events A and B in the same sample space,

$$P(A \mid B) = P(AB) / P(B).$$

This says that the probability of event A happening given that event B has happened is equal to the probability of the event A intersect B divided by the probability of event B.

The conditional probability $P(A \mid B)$ is sometimes called a **posterior probability** because it gives the probability of event A *after* the event B has already happened. The probability $P(A)$ is called a **prior probability** because it gives us the probability of event A *before* anything else has happened.

The probability of the intersection of two events, $P(AB)$, is often called the **joint probability** of the two events.

Examples of Conditional Probability

- If event B is a subset of event A, and if you know that B has happened, then A must have happened. In symbols, if $B \subset A$, then $AB = A \cap B = B$ and $P(A \mid B) = P(AB) / P(B) = P(B) / P(B) = 1$.
- If A and B are mutually exclusive events, then $P(AB) = 0$. If we know that event B has happened, it must be that the other event A *cannot* happen. In symbols, $P(A \mid B) = P(AB) / P(B) = 0 / P(B) = 0$.
- You toss one die. If the number shown is greater than three, the probability of getting an even number is P(even | greater than three) = P(even and greater than three) / P(greater than three) = $(2 / 6) / (3 / 6) = 2 / 3 \approx 0.667$.
- You toss a pair of dice. If you get a sum less than four, the probability of a two on the first die is P(two on first die | sum less than four) = P(two on the first die and sum less than four) / P(sum less than four) = $(1 / 36) / (3 / 36) = 1 / 3 \approx 0.333$. There is only one way to get two on the first die and a sum less than four, (2, 1). There are only three ways to get a sum less than four, (1, 1), (1, 2), and (2, 1).
- You toss a pair of dice. If you get a sum less than four, the probability that two appears on at least one of the dice is P(two on at least one die | sum is less than four) = P(two on at least one die and the sum is less than four | sum is less than four) = $(2 / 36) / (3 / 36) = 2 / 3 \approx 0.667$. Like the example above, you must count the outcomes. There are two ways to get two on at least one die and a sum less than four, (1, 2) and (2, 1). There are three ways to get a sum less than four, (1, 1), (1, 2), (2, 1).
- You choose one card from a deck of 52. You have been told that it is an even-numbered red card. The probability that it is less than seven is P(less than seven | even-numbered red card) = P(less than seven and even-numbered red card) / P(even-numbered red card) = $(6 / 52) / (10 / 52) = 0.6$.

- An urn contains 30 marbles; 15 are red, 10 are white, and 5 are blue. You choose one marble at random. Let R be the event that you get a red marble and W be the event that you get a white marble. You choose a marble at random and are told that it is not white. The probability that it is red is $P(R \mid W^C) = P(RW^C) / P(W^C) = P(R) / P(W^C) = (15 / 30) / (20 / 30) = 0.75$. Recall that in W^C, the superscript means the complement of the event, so RW^C is the event that the marble is red and not white, which is the same as the event that the marble is red.
- An urn contains 10 marbles; 8 are red and 2 are white. You choose two marbles at random. Let R_1 be the event that the first marble is red, and let R_2 be the event that the second marble is red. Suppose the first is red. The probability that the second marble is red is $P(R_2 \mid R_1) = P(R_2R_1) / P(R_1) = ({}_8P_2 / {}_{10}P_2) / (8 / 10) = (56 / 90) / (8 / 10) \approx 0.622 / 0.8 \approx 0.778$.
- In a school, 25% of the students play soccer, 30% play tennis, and 15% play both soccer and tennis.
  - If you know that Josie plays soccer, what is the probability that she plays tennis also? Solution: Let S be the event that a student plays soccer, and let T be the event that a student plays tennis. $P(T \mid S) = P(TS) / P(S) = 0.15 / 0.25 = 0.6$. So the probability that Josie also plays tennis is 60%.
  - If you know that Skyler plays tennis, what is the probability that he plays soccer? Solution: $P(S \mid T) = P(TS) / P(T) = 0.15 / 0.30 = 0.5$. The probability that Skyler also plays soccer is 50%.

Laboratory tests to determine the presence of a disease usually give some false positives (indication of the disease when an individual does not have the disease) and some false negatives (indication of no disease when an individual does have the disease). Doctors do not know the specific reasons why such results happen, so they treat false positives and false negatives as random events.

The developers of a diagnostic test measure the rates of false positives and false negatives empirically by counting the number of false positives and false negatives on a large population of people who have been tested. These rates are given to doctors who use the test. The doctors must then evaluate the test outcome of an individual patient based on these rates. A physician who has reason to suspect that an individual may show a false positive or a false negative result will recommend additional tests and evaluation.

Let D be the event of having the disease being tested for. Let T be the event that the test is positive. The probability $P(D)$ that an individual has the disease is determined by the rate of the disease in the population. The probability $P(T)$ that an individual tests positive is determined empirically, based on data from testing large numbers of people.

The rate of false positives is the probability an individual tests positive and does not have the disease; this is $P(TD^C)$. The rate of false negatives is the probability that an individual tests negative and does have the disease; this is $P(T^CD)$.

A doctor needs to know the likelihood that the patient does not have the disease even though the test is positive, $P(D^C \mid T)$, and the likelihood that the patient has the disease even though the test is negative, $P(D \mid T^C)$. The Example of False Positives shows how to get such information from the results of a diagnostic test.

Example of False Positives

For a given disease, the rate of false positives is 0.03. Of the general population, 6% test positive for the disease. Sawyer is tested and the result is positive. What is the probability that he does not have the disease?

Solution: We are given $P(T) = 0.06$. The rate of false positives is $P(TD^C)$, which is given as 0.03. The probability that Sawyer does not have the disease even though he tested positive is $P(D^C \mid T) = P(TD^C) / P(T) = 0.03 / 0.06 = 0.5$. This rate is high because, even though the rate of false positives is low, the rate of this disease in the general population is also low. Most people who get tested for the disease do not have it, so there will be many false positives.

6.2 Multiplication Theorem for Conditional Probability

The formula for conditional probability can be rewritten as the **Multiplication Theorem for Conditional Probability**:

$$P(AB) = P(A \mid B)\,P(B).$$

This says that the probability of the intersection of two events A and B can be computed as the product of probability of A given that B has happened times the probability of B.

The Multiplication Theorem is useful to compute the probability of an intersection of two events when we already know a conditional probability. It is sometimes called the **Theorem of Compound Probabilities**.

Example Using the Multiplication Theorem for Conditional Probability

Lucas and Morgan are going to lunch together after class. They like pizza, but the chance that they will have to wait in line at the pizza shop is 0.6. If they go to the cafeteria, they won't have to wait. They flip a coin to decide where to go. What is the chance that they will end up waiting in line for pizza?

Solution: Let Z be the event that they go for pizza, and let W be the event that they wait. $P(ZW) = P(Z)\,P(W \mid Z) = 0.5 \times 0.6 = 0.30$. This means that they have a 30% chance of waiting in line for pizza.

6.3 Switching Conditions

Suppose that you toss two dice and want to get a sum of 6. There are five ways to get a sum of 6, (1, 5), (2, 4), (3, 3), (4, 2), and (5, 1). The probability of getting a sum of 6 is $5 / 36 \approx 0.139$. If there is additional information, and you know that the first die is odd, the probability of getting a sum of 6 is greater: P(sum is 6 | first die is odd) = P(sum is 6 and first die is odd) / P(first die is odd) = $(3 / 36) / (18 / 36) \approx 0.167$.

Now look at this the other way. If you know that the sum is 6, the probability of getting an odd number is greater: the conditional probability is P(first die is odd | sum is 6) = $(3 / 36) / (5 / 36) = 0.6$.

In these two examples, we have the same events, but in the first case, the condition is that the first die is odd. In the second example, the condition is that the sum is 6. Let's look at the formulas for these two conditional probabilities.

$$P(A \mid B) = P(AB) / P(B)$$

$$P(B \mid A) = P(AB) / P(A)$$

Solving both equations for P(*AB*) we get

$$P(AB) = P(A \mid B)\, P(B)$$

$$P(AB) = P(B \mid A)\, P(A)$$

Combining these two equations, we get $P(A \mid B)\, P(B) = P(B \mid A)\, P(A)$ or

$$P(B \mid A) = \frac{P(A \mid B)\, P(B)}{P(A)}$$

This says that the conditional probability of event *B* happening given that event *A* has happened can be computed from the conditional probability of event *A* given that event *B* has happened. This formula is useful when we want to change the events in a conditional probability.

Examples of Switching Conditions

- Ellie burns CDs of her band's music. She uses Brand X, which she finds to be 1% defective, and Brand Y, which she finds to be 5% defective. She recently burned 400 CDs, 100 of Brand X and 300 of Brand Y. A friend brought back one of the CDs and said that it was defective. What is the probability it is Brand X? Brand Y?

  Solution: Let *X* be the event that the CD is Brand X, let *Y* be the event that the CD is Brand Y, and let *D* be the event that the CD is defective. Then $P(X) = 100 / 400 = 0.25$, $P(Y) = 300 / 400 = 0.75$, $P(D \mid X) = 0.01$, and $P(D \mid Y) = 0.05$. Also, since there are likely to be 16 defectives (1 of Brand X and 15 of Brand Y), the probability of a defective is $P(D) = 16 / 400 = 0.04$. We want $P(X \mid D)$ and $P(Y \mid D)$. By switching conditions, we get

$$P(X \mid D) = \frac{P(D \mid X)\,P(X)}{P(D)} = \frac{0.01 \times 0.25}{0.04} = 0.0625$$

and

$$P(Y \mid D) = \frac{P(D \mid Y)\,P(Y)}{P(D)} = \frac{0.05 \times 0.75}{0.04} = 0.9375$$

So the probability that the defective CD is Brand X is 0.0625 and that it is Brand Y is 0.9375. This probability of the defective CD being of Brand Y is high because Brand Y has a higher rate of defectives and there were more CDs of Brand Y. Note that these two probabilities add up to 1.

- Geneticists often study fruit flies because they are easy to breed and have easily identifiable traits. Jace has been breeding fruit flies and now has 100 flies that have brown or red eyes and black or yellow bodies. He has found that 30% of the flies have black bodies and 60% have brown eyes. Of the ones with brown eyes, 40% have black bodies. He chooses a fruit fly at random, and it has a black body. How likely is it to have brown eyes?

  Solution: Let N be the event that the eyes are brown, K be the event that the body is black. The data tells us that $P(N) = 0.6$, $P(K) = 0.3$, and $P(K \mid N) = 0.4$. Then

$$P(N \mid K) = \frac{P(K \mid N) \times P(N)}{P(K)} = \frac{0.4 \times 0.6}{0.3} = 0.80$$

  So the probability that a randomly chosen fruit fly with a black body will have brown eyes is 0.80.

6.4 Partitions and the Law of Total Probability

In many of the examples we have seen so far, the sample space consists of several different events. For example, with the CDs, there are two possibilities, Brand X and Brand Y. The principles of set theory can be applied to give insights into probabilities when the sample space can be written as the union of several mutually disjoint events.

Remember that two events are disjoint if their intersection is empty; several events are mutually disjoint if any two taken at a time are disjoint.

Suppose we have a sample space S that can be broken up into n different events $S_1, S_2, \ldots, S_n$ with the properties that any two are mutually disjoint, and the union of all of the events is the whole sample space, so $S = S_1 \cup S_2 \cup S_3 \cup \cdots \cup S_n$. This is a **partition** of S.

We can write any event A in the sample space S in terms of the partition like this:

$$A = AS = A \cap S = A \cap (S_1 \cup S_2 \cup S_3 \cup \cdots \cup S_n)$$
$$= (A \cap S_1) \cup (A \cap S_2) \cup (A \cap S_3) \cup \cdots \cup (A \cap S_n)$$

The events $A \cap S_1, A \cap S_2, A \cap S_3, \ldots, A \cap S_n$ are mutually disjoint and form a partition of A. The probability of A is therefore

$$\mathrm{P}(A) = \mathrm{P}(AS) = \mathrm{P}(AS_1) + \mathrm{P}(AS_2) + \mathrm{P}(AS_3) + \cdots + \mathrm{P}(AS_n)$$

Using the Multiplication Theorem for Conditional Probability for $\mathrm{P}(AS_1)$, we get

$$\mathrm{P}(AS_1) = \mathrm{P}(A \mid S_1)\ \mathrm{P}(S_1)$$

Do this for each of the probabilities of the form $\mathrm{P}(AS_i)$, getting

$$\mathrm{P}(A) = \mathrm{P}(A \mid S_1)\ \mathrm{P}(S_1) + \mathrm{P}(A \mid S_2)\ \mathrm{P}(S_2) + \mathrm{P}(A \mid S_3)\ \mathrm{P}(S_3) + \cdots + \mathrm{P}(A \mid S_n)\ \mathrm{P}(S_n)$$

We use the special summation symbol Σ to write this as

$$\mathrm{P}(A) = \sum_1^n \mathrm{P}(A \mid S_i)\ \mathrm{P}(S_i)$$

This says that for a partition of the sample space, we can compute the probability of an event if we know conditional probabilities for each of the events in the partition.

This formula is called the **Law of Total Probability**, because it takes into account the totality of all possible outcomes, or the **Law of Alternatives**, because it takes into account all possible alternative outcomes. This law is useful when it is easier to measure the conditional probabilities $\mathrm{P}(A \mid S_i)$ than it is to measure the probability of event A.

If the sample space S is a population, writing $S = S_1 \cup S_2 \cup S_3 \cup \cdots \cup S_n$ makes S into a **stratified population**, where the subsets S_i are the layers, or *strata*, of the population. A researcher might stratify a population by age, gender, race, profession, nationality, income, or some other characteristic relevant to a research study.

Examples Using the Law of Total Probability

- A school is made up of 40% seniors, 40% juniors, and 20% sophomores. Each class is polled about their preferences of activities for a school outing. Of the seniors, 40% choose kayaking, 30% choose swimming, and 30% choose hiking. Of the juniors, 30% choose kayaking, 50% choose swimming, and 20% choose hiking. Of the sophomores, 60% choose kayaking, 10% choose swimming, and 30% choose hiking. Using the Law of Total Probability, we can find the probability that a student chosen at random will prefer each of the activities. Let S represent seniors, J represent juniors, and let M represent sophomores; this gives a partition of the students. For the activities, let K represent kayaking, W represent swimming, and H hiking.

$$\begin{aligned} P(K) &= P(K \mid S)\,P(S) + P(K \mid J)\,P(J) + P(K \mid M)\,P(M) \\ &= (0.4 \times 0.4) + (0.3 \times 0.4) + (0.6 \times 0.2) = 0.40 \end{aligned}$$

$$\begin{aligned} P(W) &= P(W \mid S)\,P(S) + P(W \mid J)\,P(J) + P(W \mid M)\,P(M) \\ &= (0.3 \times 0.4) + (0.5 \times 0.4) + (0.1 \times 0.2) = 0.34 \end{aligned}$$

$$\begin{aligned} P(H) &= P(H \mid S)\,P(S) + P(H \mid J)\,P(J) + P(H \mid M)\,P(M) \\ &= (0.3 \times 0.4) + (0.2 \times 0.4) + (0.3 \times 0.2) = 0.26 \end{aligned}$$

A student chosen at random will prefer kayaking with probability 0.4, swimming with probability 0.34, and hiking with probability 0.26.

- In the same school, with 40% seniors, 40% juniors, and 20% sophomores, each class is asked about their preferences for lunch. Of the seniors, 30% choose pizza and 70% choose tacos. Of the juniors, 50% choose pizza and 50% choose tacos. Of the sophomores, 70% choose pizza and 30% choose tacos. To find the probability that a student chosen at random will prefer pizza, start by letting S represent seniors, J represent juniors, and let M represent sophomores. Let Z represent pizza and T represent tacos.

$$\begin{aligned} P(Z) &= P(Z \mid S)\,P(S) + P(Z \mid J)\,P(J) + P(Z \mid M)\,P(M) \\ &= (0.3 \times 0.4) + (0.5 \times 0.4) + (0.7 \times 0.2) = 0.46 \end{aligned}$$

This tells us that a student chosen at random will prefer pizza with probability 0.46.

6.5 Bayes' Formula for Conditional Probability

Bayes' formula gives us a way to test a hypothesis using conditional probabilities. A hypothesis is a suggested explanation for a specific outcome. If we see that a probability $P(A \mid B)$ is high, we might hypothesize that event B is a cause of the event A. We use Bayes' formula when we know conditional probabilities of the form $P(B \mid A)$ and want a conditional probability of the form $P(A \mid B)$.

For example, a botanist notices that a certain species of grass is dying in a nature preserve. She also notices that a lot of the dying grass has a type of fungus on it. Is the fungus killing the grass, or is the fungus just as likely to be found on healthy grass? The botanist can measure the probabilities that dying grass has fungus and that healthy grass has fungus. If the event of dying grass is represented by D and the event of having fungus is represented by F, then the botanist can measure $P(F \mid D)$—the probability that dying grass has fungus—and $P(F \mid H)$—the probability that healthy grass has fungus. Her hypothesis is that the fungus is killing the grass, and this means the probability that the grass dies given that it has been attacked by fungus, $P(D \mid F)$, would be high.

In this example, we have a sample space that can be partitioned into mutually exclusive events (having fungus or not) and a specific outcome (dying) that has occurred, and we want to guess or hypothesize which event in the partition the specific outcome came from. Thus the botanist is hypothesizing that the dying grass came from grass with fungus. Bayes' formula allows us to use information that we have available to evaluate how likely our hypothesis is.

For example, Graham gets e-mails from his three friends, Amiyah, Bryce, and Cora. About 20% are from Amiyah, 50% from Bryce, and 30% from Cora. This gives a partition of the sample space consisting of all of his e-mails from the three friends. Some of the attachments they send have viruses. This is another event, e-mails with a virus. Today, Graham got a virus from one of his three friends, and he is trying to determine which friend sent the virus. In order to do this, he must know the rate at which he gets viruses from his friends. He has determined that Amiyah sends viruses with 4% of her attachments, Bryce sends viruses with 10% of his attachments, and Cora sends viruses with 6% of her attachments.

This is a situation where Bayes' formula can be very useful. We have a partition of the sample space into three events: A (e-mail from Amiyah) with $P(A) = 0.2$, B (e-mail from Bryce) with $P(B) = 0.5$, and C (e-mail from Cora) with $P(C) = 0.3$. We also have three conditional probabilities where V is the event of getting a virus with the e-mail: $P(V \mid A) = 0.04$, $P(V \mid B) = 0.1$, and $P(V \mid C) = 0.06$.

Furthermore, we can use the Law of Total Probability to get $P(V) = P(VA) + P(VB) + P(VC) = (0.2 \times 0.04) + (0.5 \times 0.1) + (0.3 \times 0.06) = 0.008 + 0.05 + 0.018 = 0.076$.

We want to compute the probabilities $P(A \mid V)$, $P(B \mid V)$, and $P(C \mid V)$ to help guess whether the virus came from Amiyah, Bryce, or Cora.

The formula for switching conditions says

$$P(B \mid A) = \frac{P(A \mid B)\, P(B)}{P(A)}$$

Using this formula, we get

$$P(A \mid V) = P(V \mid A)\, P(A) / P(V) = (0.04 \times 0.2) / 0.076 = 0.105$$
$$P(B \mid V) = P(V \mid B)\, P(B) / P(V) = (0.1 \times 0.5) / 0.076 = 0.658$$
$$P(C \mid V) = P(V \mid C)\, P(C) / P(V) = (0.06 \times 0.3) / 0.076 = 0.237$$

These probabilities give numbers to the three possibilities, that Amiyah, Bryce, or Cora sent the virus. We see that it is much more likely that Bryce sent the virus.

This example shows the value of Bayes' formula, which computes the conditional probability for a guess or hypothesis about a situation given a partition of the sample space and conditional probabilities with a different order of the events.

To get Bayes' formula, start by supposing that the sample space S is the union of n different mutually disjoint events:

$$S = S_1 \cup S_2 \cup S_3 \cup \cdots \cup S_n.$$

Suppose also that we know the probabilities of each of the events S_i and all of the conditional probabilities $P(A \mid S_i)$ for some event A. The goal is to compute the conditional probability $P(S_j \mid A)$ for one of the mutually disjoint events S_j.

To get Bayes' formula, start with the formula for conditional probability:

$$P(S_j \mid A) = \frac{P(AS_j)}{P(A)}$$

Make two substitutions in this formula. First, replace $P(AS_j)$ with $P(A \mid S_i)\, P(S_j)$ in the numerator using the Multiplication Theorem for Conditional Probability. Then, use the Law of Total Probability to replace P(A) in the denominator with

$$P(A \mid S_1)\, P(S_1) + P(A \mid S_2)\, P(S_2) + P(A \mid S_3)\, P(S_3) + \cdots + P(A \mid S_n)\, P(S_n).$$

This gives us Bayes' formula:

$$P(S_j \mid A) = \frac{P(A \mid S_j)\, P(S_j)}{P(A \mid S_1)\, P(S_1) + P(A \mid S_2)\, P(S_2) + P(A \mid S_3)\, P(S_3) + \cdots + P(A \mid S_n)\, P(S_n)}$$

or, using summation notation,

$$P(S_j \mid A) = \frac{P(A \mid S_j)\, P(S_j)}{\sum_1^n P(A \mid S_i)\, P(S_i)}.$$

We use Bayes' formula with inference or hypothesis testing, when a random choice is made from a stratified population. If a random choice gives outcome A, we might hypothesize that it came from subpopulation S_j. The probability that this guess is correct is $P(S_j \mid A)$ and can be computed using Bayes' formula. While this doesn't tell us for sure where A came from, it can help evaluate how good our hypothesis is given the data that we have available.

Let's consider an example with marbles in urns because it can be used as a model for many other situations. Suppose there are three urns containing marbles as follows:

> First urn: 5 red marbles, 8 yellow marbles, and 7 green marbles (20 marbles)
>
> Second urn: 10 red marbles and 10 yellow marbles (20 marbles)
>
> Third urn: 3 red marbles, 2 yellow marbles, and 5 green marbles (10 marbles)

Someone has chosen an urn at random and drawn one of the marbles at random from that urn and has given it to you. You know that the marble is green, and you want to know which urn it came from. Use Bayes' formula to get probabilities that can help you.

Let S_1, S_2, and S_3 be the events that the marble came from the first, second, or third urn, respectively. Let R, Y, and G be the events of getting a red, yellow, or green marble.

You can guess that it came from urn S_1 and try to find the probability that it came from there. This guess is the hypothesis. The probability that this hypothesis is correct is the conditional probability $P(S_1 \mid G)$, the probability that it came from the first urn given that it is green.

$$P(S_1 \mid G) = \frac{P(G \mid S_1)\, P(S_1)}{P(G \mid S_1)\, P(S_1) + P(G \mid S_2)\, P(S_2) + P(G \mid S_3)\, P(S_3)}$$

$$= \frac{(7/20)\times(1/3)}{(7/20)\times(1/3)+(0/20)\times(1/3)+(5/10)\times(1/3)} \approx 0.412$$

Similarly, for the second urn (though we already know that a green marble could not have come from that one),

$$P(S_2 \mid G) = \frac{P(G \mid S_2)\,P(S_2)}{P(G \mid S_1)\,P(S_1) + P(G \mid S_2)\,P(S_2) + P(G \mid S_3)\,P(S_3)}$$

$$= \frac{(0/20) \times (1/3)}{(7/20) \times (1/3) + (0/20) \times (1/3) + (5/10) \times (1/3)} = 0$$

Finally, for the third urn, we get

$$P(S_3 \mid G) = \frac{P(G \mid S_3)\,P(S_3)}{P(G \mid S_1)\,P(S_1) + P(G \mid S_2)\,P(S_2) + P(G \mid S_3)P(S_3)}$$

$$= \frac{(5/10) \times (1/3)}{(7/20) \times (1/3) + (0/20) \times (1/3) + (5/10) \times (1/3)} \approx 0.588$$

From these answers, we see that the green marble is a little more likely to have come from the third urn than the first urn; it could not have been from the second urn.

Note that all of the denominators are the same, so we only need one computation for the denominator. Also, all of the probabilities should add up to 1, as they do here. In other cases, the sum may not be 1 but only close to 1 due to rounding errors.

Examples Using Bayes' Formula

- Three production lines produce printer cartridges. One produces 60% of output, another produces 30% of the output, and the third produces 10% of the output. Testing has shown that these production lines average 2%, 5%, and 3% defective output, respectively. A cartridge is chosen at random from the day's output and is found defective. Which production line did it come from?

Solution: We can let A, B, and C stand for the events that the random cartridge is from the first, second, or third production line. Then $P(A) = 0.6$, $P(B) = 0.3$, and $P(C) = 0.1$. Let D be the event that the cartridge is defective. The probabilities that the cartridge is defective and from first, second, or third production line, respectively, are $P(D \mid A) = 0.02$, $P(D \mid B) = 0.05$, and $P(D \mid C) = 0.03$. Using Bayes' formula, we see that

$$
\begin{aligned}
P(A \mid D) &= \frac{P(D \mid A)\,P(A)}{P(D \mid A)\,P(A) + P(D \mid B)\,P(B) + P(D \mid C)\,P(C)} \\
&= \frac{0.02 \times 0.6}{(0.02 \times 0.6) + (0.05 \times 0.3) + (0.03 \times 0.1)} \\
&= \frac{0.012}{0.012 + 0.015 + 0.003} \\
&= \frac{0.012}{0.030} = 0.40
\end{aligned}
$$

Similarly,

$$P(B \mid D) = \frac{P(D \mid B)\,P(B)}{0.030} = \frac{0.05 \times 0.3}{0.030} = \frac{0.015}{0.030} \approx 0.50$$

and

$$P(C \mid D) = \frac{P(D \mid C)\,P(C)}{0.030} = \frac{0.03 \times 0.1}{0.030} = \frac{0.003}{0.030} \approx 0.10$$

It is impossible to say which production line the defective cartridge came from, but it is most likely that it came from the second production line and least likely that it came from the third.

- Arla decides to guess on an exam if she doesn't know the correct answer. She knows the answers to 60% of the questions. The other questions are multiple choice, so she has a 25% chance to get the correct answer. Arla answered the first question correctly. What is the chance she knew the answer?

Solution: Let K be the event that she knows the answer, so $P(K) = 0.6$. Then K^C is the event that she must guess, so $P(K^C) = 0.4$. Let C be the event that she answers a question correctly. Also, $P(C \mid K) = 1$ because if she knows the answer, she will get the question correct. Then,

$$P(K \mid C) = \frac{P(C \mid K)\,P(K)}{P(C \mid K)\,P(K) + P(C \mid K^C)\,P(K^C)}$$

$$= \frac{1 \times 0.6}{(1 \times 0.6) + (0.25 \times 0.4)} = \frac{0.6}{0.7} \approx 0.857$$

So if she got the question right, the probability is 0.857 that she knew the answer and did not guess.

- There are three urns, and each urn contains two coins. The first contains two nickels, the second contains two dimes, and the third contains a nickel and a dime. An urn is chosen at random, and a coin is taken out at random. It is a nickel. What is the probability that the other coin in the urn is a nickel?

  Solution: Call the urns U_1, U_2, and U_3; then $P(U_1) = P(U_2) = P(U_3) = 1/3$. We want to know how likely it was that the nickel came from the first urn because that would mean that the other coin in the urn is a nickel. So we compute $P(U_1 \mid \text{nickel})$ using Bayes' formula.

$$P(U_1 \mid \text{nickel})$$

$$= \frac{P(\text{nickel} \mid U_1)\,P(U_1)}{P(\text{nickel} \mid U_1)\,P(U_1) + P(\text{nickel} \mid U_2)\,P(U_2) + P(\text{nickel} \mid U_3)\,P(U_3)}$$

$$= \frac{1 \times (1/3)}{1 \times (1/3) + 0 \times (1/3) + (1/2) \times (1/3)}$$

$$= \frac{1/3}{3/6} = \frac{2}{3}$$

This problem is called Russell's Paradox because you might think that once you get a nickel, you have narrowed the choice of urns to two and each is equally likely. However, even though getting a nickel narrows the choice of urns to two, the first urn was twice as likely as the third to have given the nickel.

6.6 Independent Events

We saw earlier that if we want to draw a king from a deck of cards, knowing that we have a club doesn't give us any information. This says that P(K | ♣) = P(K). Similarly, if we want to draw a club, knowing that we have a king doesn't give any information, and we have P(♣ | K) = P(♣). In such cases, where information about one event doesn't give any information about another event, we say that these two events are **independent events.**

Let's look at two different ways to compute P(K♣), which is the probability of drawing the king of clubs. The formula for conditional probability says that P(K | ♣) = P(K♣) / P(♣). But we know that P(K | ♣) is the same as P(K), so we substitute this into the conditional probability formula to get P(K) = P(K♣) / P(♣). Multiply both sides by P(♣) and get

$$P(K\clubsuit) = P(K) \times P(\clubsuit)$$

We can verify this by looking at the following numbers.

$$P(K) = 4 / 52$$

$$P(\clubsuit) = 13 / 52$$

$$P(K\clubsuit) = 1 / 52$$

We can see that P(K) × P(♣) = (4 / 52) × (13 / 52) = 52 / (52 × 52) = 1 / 52. But this is just the probability P (K♣) = 1 / 52, and we confirm that

$$P(K\clubsuit) = P(K) \times P(\clubsuit).$$

These observations lead to the definition of **independent events**:
Two events A and B in the same sample space are independent if $P(AB) = P(A)\,P(B)$.

This formula gives us a new and simpler way to characterize independent events. Two events A and B are independent if the probability of both events happening together is equal to the product of the probabilities of the two events.

There are two ways we can use this formula. If we know $P(A)$, $P(B)$, and $P(AB)$, we can see if $P(A) \times P(B) = P(AB)$. If so, then we can conclude that A and B are independent events; if not, we can conclude that they are not independent events.

On the other hand, if we know that events A and B are independent, and we know two of the three probabilities P(A), P(B), and P(AB), we can solve for the third using the equation P(A) P(B) = P(AB).

Events that are independent are also called **stochastically independent** or **statistically independent**.

Examples of Independent Events

- Throw two dice. Let E be the event that the first die is odd, and let F be the event the second die is 6. P(E) = 18 / 36 = 1 / 2, P(F) = 6 / 36 = 1 / 6. The event EF contains three outcomes, (1, 6), (3, 6), and (5, 6), so P(EF) = 3 / 36 = 1 / 12. Since

  $$P(E)\,P(F) = (1 / 2) \times (1 / 6) = 1 / 12 = P(EF),$$

  the two events E and F are independent.
- Flip three coins. The sample space is {HHH, HHT, HTH, HTT, THH, THT, TTH, TTT}. Consider three events, A, B, and C. Event A is that the last flip is H, A = {HHH, HTH, THH, TTH}. The event B is that the middle flip is H, B = {HHH, HHT, THH, THT}. The event C is that there are exactly two heads in a row, C = {HHT, THH}. P(A) = P(B) = 0.5 and P(C) = 0.25.
  - Events A and B are independent because AB is the event {HHH, THH} and P(AB) = 0.25. This is the same as P(A) × P(B) = 0.5 × 0.5 = 0.25.
  - Events A and C are independent because AC is the event {THH} and P(AC) = 1 / 8 = 0.125. This is the same as P(A) × P(C) = 0.5 × 0.25 = 0.125.
  - However, events B and C are not independent because BC is the event {HHT, THH} and P(BC) = 2 / 8 = 0.25. This is not the same as P(B) × P(C) = 0.5 × 0.25 = 0.125.

For three events A, B, and C to be independent, they must satisfy four conditions: A and B must be independent; A and C must be independent; B and C must be independent; and finally, P(ABC) = P(A) P(B) P(C). You can define independence for four or more events similarly.

More Examples of Independent Events

- Throw two dice. Consider the three events: E, that the sum is even; F, that the first die is odd; and G, that the second die is even. Are these events independent?

  Solution: $P(E) = 0.5$, $P(F) = 0.5$, and $P(G) = 0.5$. Also, $P(EF) = 0.25$, $P(EG) = 0.25$, and $P(FG) = 0.25$. Taken two at a time, the events are independent. However, $P(EFG) = 0$ and $P(E)\,P(F)\,P(G) = 0.5 \times 0.5 \times 0.5 = 0.125$. So the three events are not independent.

- Throw three dice. Consider the three events: E, that the first die is odd; F, that the second die is odd; and G, that the third die is even. Are these events independent?

  Solution: $P(E) = 0.5$, $P(F) = 0.5$, and $P(G) = 0.5$. Also, $P(EF) = 0.25$, $P(EG) = 0.25$, and $P(FG) = 0.25$. Taken two at a time, the events are independent. Also, $P(EFG) = 0.125$ and $P(E)\,P(F)\,P(G) = 0.5 \times 0.5 \times 0.5 = 0.125$. So the three events are independent.

Chapter 6 Summary

The **joint probability** of two events A and B is the probability of the intersection of the events, $P(AB)$.

The **conditional probability** for two events A and B in the same sample space, written $P(A \mid B)$, is the probability of event A happening, knowing that event B has already happened. It can be computed using the formula $P(A \mid B) = P(AB) / P(B)$.

The conditional probability $P(A \mid B)$ is a **posterior probability** because it gives the probability of event A *after* event B has already happened. The probability $P(A)$ is a **prior probability** because it gives us the probability of event A *before* anything else has happened.

A **false positive** is a test result showing the presence of a disease or condition when it is not present. A **false negative** is a test result showing no disease when it is present.

The **Multiplication Theorem for Conditional Probability** states

$$P(AB) = P(A \mid B)\, P(B).$$

The **formula for switching conditions** says

$$P(B \mid A) = \frac{P(A \mid B)\, P(B)}{P(A)}$$

A **partition** of a sample space S is a collection of events $S_1, S_2, \ldots, S_n$ such that any two are mutually disjoint and $S = S_1 \cup S_2 \cup S_3 \cup \cdots \cup S_n$.

The **Law of Total Probability** or the **Law of Alternatives** says that for an event A in the sample space S,

$$P(A) = P(A \mid S_1)\, P(S_1) + P(A \mid S_2)\, P(S_2) + P(A \mid S_3)\, P(S_3) + \cdots + P(A \mid S_n)\, P(S_n)$$

or, using summation notation,

$$P(A) = \sum_1^n P(A \mid S_i)\, P(S_i)$$

A **stratified population** is a partition of a population $S = S_1 \cup S_2 \cup S_3 \cup \cdots \cup S_n$, and the subsets S_i are the **layers** of the population.

Suppose a sample space S is the union of n different mutually disjoint events, $S = S_1 \cup S_2 \cup S_3 \cup \cdots \cup S_n$. **Bayes' formula** says

$$P(S_j \mid A) = \frac{P(A \mid S_j)\, P(S_j)}{P(A \mid S_1)\, P(S_1) + P(A \mid S_2)\, P(S_2) + P(A \mid S_3)\, P(S_3) + \cdots + P(A \mid S_n)\, P(S_n)}$$

or, using summation notation,

$$P(S_j \mid A) = \frac{P(A \mid S_j)\, P(S_j)}{\sum_1^n P(A \mid S_i)\, P(S_i)}$$

Bayes' formula is often used in inference or hypothesis testing. If event A has happened, and event S_j is the **hypothesis**, then $P(S_j \mid A)$ is the probability that this hypothesis is correct.

Two events A and B in the same sample space are **independent** (or **stochastically independent** or **statistically independent**) if $P(AB) = P(A)\, P(B)$.

Chapter 6 Practice Problems

Conditional Probability

1. You draw one card from a deck of 52.
   a. What is the probability of drawing a heart?
   b. What is the probability of drawing a heart if you know that you have drawn a red card?
   c. What is the probability of drawing a heart if you know that you have a face card?
   d. What is the probability of drawing a face card if you know that you have a heart?
2. An urn contains 40 marbles; 25 are red, 10 are white, and 5 are blue. You choose one marble at random.
   a. What is the probability of choosing a white marble?
   b. What is the probability of choosing a marble that is not blue?
   c. You choose a marble at random and are told that it is not blue. What is the probability that it is white?
   d. You choose a marble at random and are told that it is not blue. What is the probability that it is red?
3. An urn contains 12 marbles; 10 are red and 2 are white. You choose two marbles at random.
   a. What is the probability that the first marble is red?
   b. What is the probability that the first marble is white?
   c. You choose two marbles at random. What is the probability that the second marble is red if you know that the first marble is red?
   d. You choose two marbles at random. What is the probability that the second marble is red if you know that the first marble is white?
4. An insurance company finds that the probability that a homeowner has a claim in any given year is 0.0725. They find that the homeowners likely to have repeated claims make up about 15% of their policyholders and have a rate of 0.2 claims per year. What is the probability that a homeowner will have repeated claims if there is a claim in the homeowner's first year with the company?

5. Events E, F, and G belong to the same sample space. Given the probabilities $P(E) = 0.6$, $P(F) = 0.55$, $P(EF) = 0.2$, $P(G) = 0.35$, and $P(G \mid F) = 0.25$, compute the following:
   a. $P(E \mid F)$
   b. $P(F \mid E)$
   c. $P(F \mid G)$

Bayes' Formula

6. A factory producing headsets has three assembly lines, A, B, and C, which manufacture 20%, 30%, and 50% of the total production, respectively. It is known that the products from assembly lines A, B, and C are 2%, 1%, and 3% defective, respectively. A headset is chosen at random from the output of the factory and is found to be defective. What is the probability that it came from assembly lines A, B, and C?
7. An archeologist is evaluating a pottery shard. The pattern on the shard makes her 40% certain that it came from City A and 60% certain that it came from the surrounding area. Testing of the clay indicates that it is of a type used in the city all of the time but in the surrounding area only 20% of the time. How likely is it that the shard came from the city?
8. A lake contains two kinds of trout, rainbow and brown. It is estimated that 40% of the trout are rainbow. Of the rainbow trout, 60% are mature, and of the brown trout, 30% are mature. You have a trout on the line and see that it is a keeper. What is the probability that it is a rainbow?

Independent Events

9. Which of the following pairs or triples of events are independent?
   a. Drawing a card, getting a face card or getting a club.
   b. Drawing a card, getting a face card or getting an ace.
   c. Drawing a card, getting a red card or getting a queen.
   d. Drawing a card, getting a red face card or getting the jack of diamonds.
   e. Throwing two dice, getting doubles or getting an even sum.

f. Throwing two dice, getting exactly one 5 or getting an even sum.

g. Throwing two dice, getting two different numbers or getting exactly one 5.

h. Throwing two dice, getting two different numbers or getting an even sum.

i. Events E and F, where P(E) = 0.5, P(F) = 0.3, and P(EF) = 0.25.

j. Events E and F, where P(E) = 0.5, P(F) = 0.3, and P(EF) = 0.15.

k. Events E, F, and G, where P(E) = 0.5, P(F) = 0.3, P(G) = 0.25, P(EF) = 0.15, P(EG) = 0.125, P(FG) = 0.075, and P(EFG) = 0.0375.

l. Events E, F, and G, where P(E) = 0.4, P(F) = 0.8, P(G) = 0.3, P(EF) = 0.32, P(EG) = 0.12, P(FG) = 0.24, and P(EFG) = 0.086.

Answers to Chapter 6 Practice Problems

1. a. 13 / 52 = 0.25

   b. P(♥ | red card) = P(red ♥) / P(red card) = (13 / 52) / (26 / 52) = 0.5

   c. P(♥ | face card) = P(♥ face card) / P(face card) = (3 / 52) / (12 / 52) = 0.25

   d. P(face card | ♥) = P(♥ face card) / P(♥) = (3 / 52) / (13 / 52) ≈ 0.231

2. a. P(white) = 10 / 40 = 0.25

   b. P(not blue) = 35 / 40 = 0.875

   c. P(white | not blue) = P(white and not blue) / P(not blue) = (10 / 40) / (35 / 40) ≈ 0.286

   d. P(red | not blue) = P(red and not blue) / P(not blue) = (25 / 40) / (35 / 40) ≈ 0.714

3. a. P(first is red) = 10 / 12 ≈ 0.833

   b. P(first is white) = 2 / 12 ≈ 0.167

   c. P(second is red | first is red) = P(second is red and first is red) / P(first is red) = $({}_{10}P_2 / {}_{12}P_2)$ / (10 / 12) = 9 / 11 ≈ 0.818

   d. P(second is red | first is white) = P(second is red and first is white) / P(first is white) = $[(2 \times 10) / {}_{12}P_2]$ / (2 / 12) = 10 / 11 ≈ 0.909

4. The sample space is all policyholders for a given year. Let C be the event containing policyholders with a claim, so $P(C) = 0.0725$. Let R be the event containing policyholders with repeated claims, so $P(R) = 0.15$. The claim rate of this group is 0.2 claims per year, so $P(C \mid R) = 0.2$. Use the formula for switching conditions to get $P(R \mid C) = P(C \mid R)\, P(R) / P(C) = (0.2 \times 0.15) / 0.0725 \approx 0.414$.
5. **a.** $0.2 / 0.55 \approx 0.3636$

   b. $0.2 / 0.6 \approx 0.3333$

   c. $0.25 \times 0.55 / 0.35 \approx 0.3929$
6. Let events A, B, and C correspond to headsets from assembly lines A, B, and C, respectively, and let D be the event that the headset is defective. We are asked to find $P(A \mid D)$, $P(B \mid D)$, and $P(C \mid D)$. Use Bayes' formula:

   $P(A \mid D) = P(A)\, P(D \mid A) / [P(A)\, P(D \mid A) + P(B)\, P(D \mid A) + P(C)\, P(D \mid A)]$
   $= (0.2 \times 0.02) / 0.022 \approx 0.1818$

   $P(B \mid D) = P(B)\, P(D \mid A) / [P(A)\, P(D \mid A) + P(B)\, P(D \mid A) + P(C)\, P(D \mid A)]$
   $= (0.3 \times 0.01) / 0.022 \approx 0.1364$

   $P(C \mid D) = P(C)\, P(D \mid A) / [P(A)\, P(D \mid A) + P(B)\, P(D \mid A) + P(C)\, P(D \mid A)]$
   $= (0.5 \times 0.030 / 0.022 \approx 0.6818$
7. Let A be the event that the shard came from City A and let C be the event that it is of the type of clay used in City A. From the information given in the problem, we know that for the particular pottery shard, $P(A) = 0.4$ based on the pattern; $P(C \mid A) = 1$; and $P(C \mid A^C) = 0.2$. We want to know $P(A \mid C)$. Use Bayes' formula: $P(A \mid C) = P(C \mid A)\, P(A) / [P(C \mid A)\, P(A) + P(C \mid A^C)\, P(A^C)] = (1 \times 0.4) / (1 \times 0.4 + 0.2 \times 0.6) \approx 0.769$.
8. Let R be the event of catching a rainbow trout, B a brown trout, and M a mature trout (keeper). We are given $P(R) = 0.4$; $P(B) = 0.6$; and $P(M \mid R) = 0.6$. We want to know $P(R \mid M)$. Use Bayes' formula: $P(R \mid M) = P(M \mid R)\, P(R) / (P(M \mid R)\, P(R) + P(M \mid B)\, P(B)) = 0.6 \times 0.4 / (0.6 \times 0.4 + 0.3 \times 0.6) \approx 0.571$.
9. **a.** Independent. P(face card) = 12 / 52; P(♣) = 13 / 52; P(♣ face card) = 3 / 52; and (12 / 52) × (13 / 52) = 3 / 52.

   b. Not independent. P(face card) = 12 / 52; P(A) = 4 / 52; P(face card and A) = 0

   c. Independent. P(red card) = 1 / 2; P(Q) = 4 / 52 = 1/ 13; P(red Q) = 2 / 52 = 1 / 26; and (1 / 2) × (1 / 13) = 1 / 26.

d. Not independent. P(red face card) = 6 / 52 = 3 / 26; P(J♦) = 1 / 52; but P(red face card and J♦) = P(J♦) = 1 / 52 ≠ (3/ 26) × (1 / 52).

e. Not independent. P(doubles) = 6 / 36 = 1 / 6; P(even sum) = 18 / 36 = 1 / 2; but P(doubles and even sum) = P(doubles) = 6 / 36 ≠ (1 / 6) × (1 / 2).

f. Not independent. P(one 5) = 10 / 36; P(even sum) = 1 / 2; P(one 5 and even sum) = 4 / 36 = 1 / 9; but (10 / 36) × (1 / 2) = 5 / 36 ≠ (1 / 2) × (1 / 9).

g. Not independent. P(two different numbers) = 30 / 36 = 5 / 6; P(one 5) = 10 / 36 = 5 / 18; but P(two different numbers and one 5) = P(one 5) = 10 / 36 = 5 / 18 ≠ (5 / 6) × (5 / 18).

h. Not independent. P(two different numbers) = 30 / 36 = 5 / 6; P(even sum) = 1 / 2; P(two different numbers and an even sum) = 12 / 36 = 1 / 3 ≠ (5 / 6) × (1 / 2).

i. Not independent. P(E) × P(F) = 0.5 × 0.3 = 0.15 ≠ P(EF) = 0.25.

j. Independent. P(E) × P(F) = 0.5 × 0.3 = 0.15 = P(EF) = 0.15.

k. Independent. P(E) P(F) = 0.15 = P(EF); P(E)P(G) = 0.125 = P(EG); P(F)P(G) = 0.75 = P(FG); and P(E) P(F) P(G) = 0.0375 = P(EFG).

l. Not independent. P(E) P(F) = 0.32 = P(EF); P(E)P(G) = 0.24 = P(EG); P(F)P(G) = 0.24 = P(FG); but P(E) P(F) P(G) = 0.096 ≠ P(EFG).

Chapter 7

Discrete Random Variables

Sometimes when random events are involved, we are more interested in something that depends on the outcome of a random trial than we are in the outcome itself. For example, we are more interested in the sum of the two dice rather than in what numbers give the sum (unless, of course, we get doubles) when playing a board game. When someone wins a lottery, their friends are more interested in the value of the prize than what numbers came up.

Random variables give a number, like the sum of two dice, that depends on a random outcome, like a toss of two dice. We use random variables when we want to look at the probability that a certain number will occur rather than the probability of an event that gives the number.

In this chapter, we will look at **discrete random variables**—that means random variables that take on only integer values or only a few values.

There are many different random variables that we will look at: Bernoulli, binomial, Poisson, geometric, negative geometric, hypergeometric, and multinomial. For this chapter, you will need to understand the concepts about set theory given in Chapter 3, the counting techniques given in Chapter 4, and the methods of computing probabilities given in Chapter 5.

7.1 Random Variables

For a roll of two dice, the sample space contains 36 different outcomes. If we look at the **sum** of the two dice, we get eleven different sums, 2, 3, 4, 5, 6, 7, 8, 9, 10, 11, and 12. The sum depends on the outcome, and different outcomes can have the same sum. This is one example of a random variable, which assigns a number to each element of a sample space.

A random variable is not a variable like the variables x and y of algebra that represent unknown quantities. In probability, a **random variable** is a rule or function that assigns a number to each element of a sample space. In other words, a random variable gives a number for each outcome of a random experiment.

A random variable can also be called a **stochastic variable** or **variate**.

Random variables are often denoted by capital letters such as X or Y. In the preceding example, if we call the random variable that gives the sum of two dice X, then we have $X(2, 5) = 7$, $X(4, 1) = 5$, and so on. The expression $X(2, 5) = 7$ tells us that the random variable X assigns the number 7 to the element in the sample space (2, 5).

For the same sample space, we could have another random variable Y that assigns to the throw of two dice the lower number shown. Thus $Y(2, 5) = 2$ and $Y(4, 1) = 1$.

The set of different values that a random variable can take on is called the **range** or the **range space** of the random variable. The range space of the sum random variable X previously given is $\{2, 3, 4, 5, 6, 7, 8, 9, 10, 11, 12\}$, and the range space for the lower number random variable Y previously given is $\{1, 2, 3, 4, 5, 6\}$.

Examples of Random Variables

- Toss four coins. The random variable X that gives the number of heads has range $\{1, 2, 3, 4\}$ with $X(\text{HHTH}) = 3$, $X(\text{THTH}) = 2$, and so on.
- Throw three dice. The random variable Y that gives the sum of the numbers has range $\{3, 4, 5, \ldots, 16, 17, 18)$ with $Y(3, 1, 6) = 10$, $Y(5, 5, 2) = 12$, and so on.

7.2 Probability Frequency Function

Once we have a random variable, we can determine the probability that the random variable will have a certain value. Since we are more interested in the value of the random variable than the outcomes that produced it, we will also be more interested in the probability that the random variable will take on a certain value than the probabilities of the outcomes that give it.

In the preceding example, where X is the random variable giving the sum of two dice, we compute the probability that the random variable will take on a certain value by counting how many outcomes give the value. This creates a new event, the set of outcomes that have a certain value of the random variable. Then we find the probability of that new event.

For throwing two dice, let's look at the event $X = 6$ as an example. Five outcomes of the 36 possible outcomes give a total of 6, namely, (1, 5), (2, 4), (3, 3), (4, 2), and (5, 1). The probability that X is 6 is therefore $5 / 36 \approx 0.139$. We write this as $P\{X = 6\} \approx 0.139$ or $P(X = 6) \approx 0.139$. Sometimes, the Greek letter rho ρ (with a subscript for the random variable) is used, so $\rho_X(6) \approx 0.139$.

The function that assigns a probability to each number in the range of a random variable is called the **probability frequency function**, or sometimes simply the **probability function** or the **probability mass function**.

For this example, all values of the probability frequency function are listed here:

$$
\begin{aligned}
P\{X = 2\} &= 1 / 36 \approx 0.0278 \\
P\{X = 3\} &= 2 / 36 \approx 0.0556 \\
P\{X = 4\} &= 3 / 36 \approx 0.0833 \\
P\{X = 5\} &= 4 / 36 \approx 0.111 \\
P\{X = 6\} &= 5 / 36 \approx 0.139 \\
P\{X = 7\} &= 6 / 36 \approx 0.167 \\
P\{X = 8\} &= 5 / 36 \approx 0.139 \\
P\{X = 9\} &= 4 / 36 \approx 0.111 \\
P\{X = 10\} &= 3 / 36 \approx 0.0833 \\
P\{X = 11\} &= 2 / 36 \approx 0.0556 \\
P\{X = 12\} &= 1 / 36 \approx 0.0278
\end{aligned}
$$

Note that these probabilities should sum to 1 because we have listed the probabilities for all possible values of the random variable. In this case, however, they add up to 1.0004 as a result of rounding errors.

A graph is a convenient way to represent a probability function. For each possible value of the random variable on the x-axis, the probability of getting that value is shown in the vertical or y-direction. Vertical lines are drawn to make the graph easier to read. Figure 7.1 shows the probability function of the random variable X that gives the sum of two dice. Since the values of probabilities are small, the vertical axis has been stretched for greater clarity.

The graph makes it easy to see the important features of a probability function. For this example, a sum equal to 7 is the most likely outcome. The graph has symmetry, so sums 2 and 12 have the same probability as do sums 3 and 11, 4 and 10, 5 and 9, and 6 and 8.

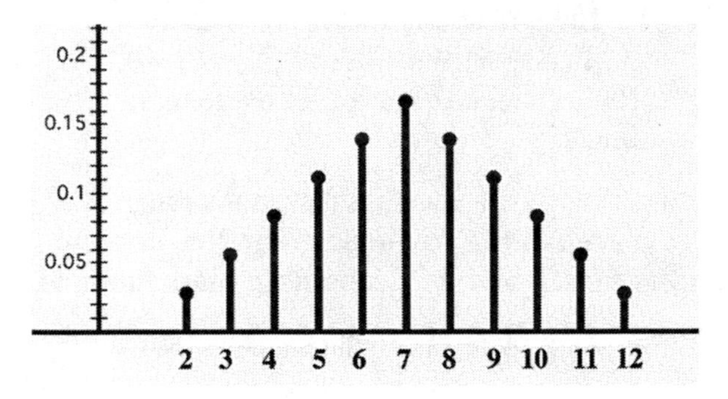

Figure 7.1
The probability frequency function of the random variable X that gives the sum of two dice.

For the random variable Y that gives the lower value of the two dice, the values of the probability function are

$$\begin{aligned}
P\{Y = 1\} &= 11 / 36 \approx 0.306 \\
P\{Y = 2\} &= 9 / 36 = 0.250 \\
P\{Y = 3\} &= 7 / 36 \approx 0.194 \\
P\{Y = 4\} &= 5 / 36 \approx 0.139 \\
P\{Y = 5\} &= 3 / 36 \approx 0.0833 \\
P\{Y = 6\} &= 1 / 36 \approx 0.0278
\end{aligned}$$

Two ways to look at the probability function for Y are shown in Figure 7.2. The one on the right uses bars the same way that a histogram does.

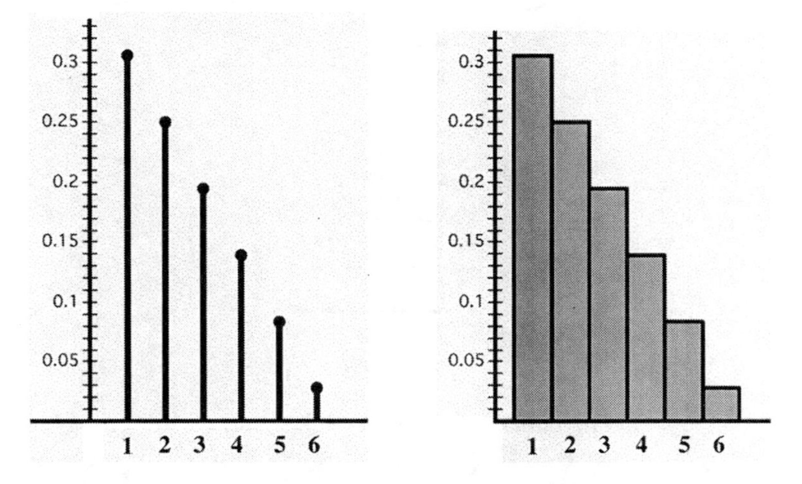

Figure 7.2
The two ways to view the probability frequency function of the random variable *Y* that gives the lower value of two dice.

Examples of Probability Frequency Functions

- Toss four coins. The random variable Z gives the number of heads.

$$P\{Z = 0\} = 1 / 16 = 0.0625$$
$$P\{Z = 1\} = 4 / 16 = 0.25$$
$$P\{Z = 2\} = 6 / 16 = 0.375$$
$$P\{Z = 3\} = 4 / 16 = 0.25$$
$$P\{Z = 4\} = 1 / 16 = 0.0625$$

The graph of this probability function is shown in Figure 7.3.

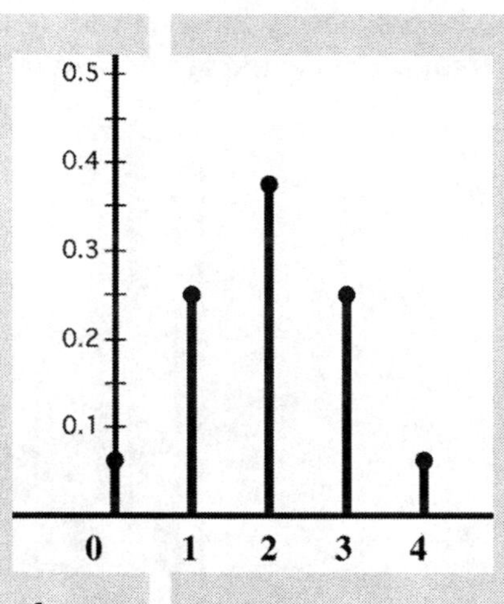

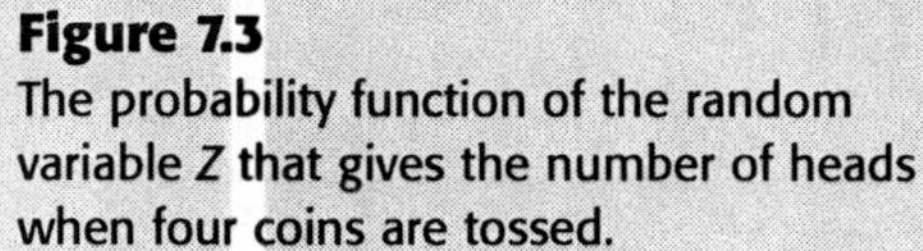

Figure 7.3
The probability function of the random variable Z that gives the number of heads when four coins are tossed.

- Throw three dice. The random variable W gives the sum of the numbers.

$$P\{W = 3\} = 1/216 \approx 0.00463$$
$$P\{W = 4\} = 3/216 \approx 0.0139$$
$$P\{W = 5\} = 6/216 \approx 0.0278$$
$$P\{W = 6\} = 10/216 \approx 0.0463$$

$$\vdots$$

$$P\{W = 16\} = 6/216 \approx 0.0278$$
$$P\{W = 17\} = 3/216 \approx 0.0139$$
$$P\{W = 18\} = 1/216 \approx 0.00463$$

Sometimes, you may be interested in the probability that a random variable takes on a value that lies in an interval. So for this last example,

$$P\{W \geq 16\} = (6 + 3 + 1) / 216 \approx 0.0463$$

and

$$P\{4 \leq W \leq 6\} = (3 + 6 + 10) / 216 \approx 0.0880$$

Probability frequency functions have several important properties. Let X be a random variable.

- All probabilities have values in the interval from 0 to 1, so

$$0 \leq P\{X = a\} \leq 1$$

- The sum of the probabilities for all outcomes in any sample space is 1. Use the summation symbol and sum over all possible values that the random variable can have:

$$\sum_a P\{X = a\} = 1$$

- The outcomes $X \leq a$ and $X > a$ are complementary events for any number a. Complementary events have probabilities that add up to 1. This means that

$$P\{X \leq a\} = 1 - P\{X > a\}$$

7.3 Cumulative Distribution Function of a Random Variable

The probability frequency function gives the probabilities for each value in the range of a random variable. The cumulative distribution function of a random variable is a different way to help us understand the properties of a random variable. For a given value x of the random variable, the **cumulative distribution function** gives the probability of the random variable taking on a value up to and including the given value x. Using symbols, we write the cumulative distribution function as the function $F(x) = P\{X \leq x\}$.

The cumulative distribution function is also called the **CDF**, or **probability distribution** or **distribution function.**

For example, the cumulative distribution function F for the random variable X that gives the number of heads takes on the value 1 / 36 when x is 2, because there is only one outcome of the random variable up to and including 2. That happens when both dice come up 1.

When x is 3, there are three outcomes with a value up to and including 3 for the random variable, namely (1, 1), (1, 2), and (2, 1), so $F(3) = 3 / 36$. All values for the CDF F are given here:

$$
\begin{aligned}
F(2) &= 1/36 \approx 0.0278 \\
F(3) &= 3/36 \approx 0.0833 \\
F(4) &= 6/36 \approx 0.167 \\
F(5) &= 10/36 \approx 0.278 \\
F(6) &= 15/36 \approx 0.417 \\
F(7) &= 21/36 \approx 0.583 \\
F(8) &= 26/36 \approx 0.722 \\
F(9) &= 30/36 \approx 0.833 \\
F(10) &= 33/36 \approx 0.917 \\
F(11) &= 35/36 \approx 0.972 \\
F(12) &= 36/36 = 1.000
\end{aligned}
$$

The graph of a cumulative distribution function shows how the probability increases as greater and greater values of the random variable are included. The graph of the CDF F is shown in Figure 7.4.

Cumulative distribution functions have several important properties. Let $F(x)$ be the cumulative distribution function of a random variable X.

- The values of the cumulative distribution function $F(x)$ keep getting larger; in fact, this is why it is called the *cumulative* distribution function. So for numbers a and b,

  $$F(a) \leq F(b) \text{ if } a < b.$$

  We also say $F(x)$ is a **nondecreasing** function.

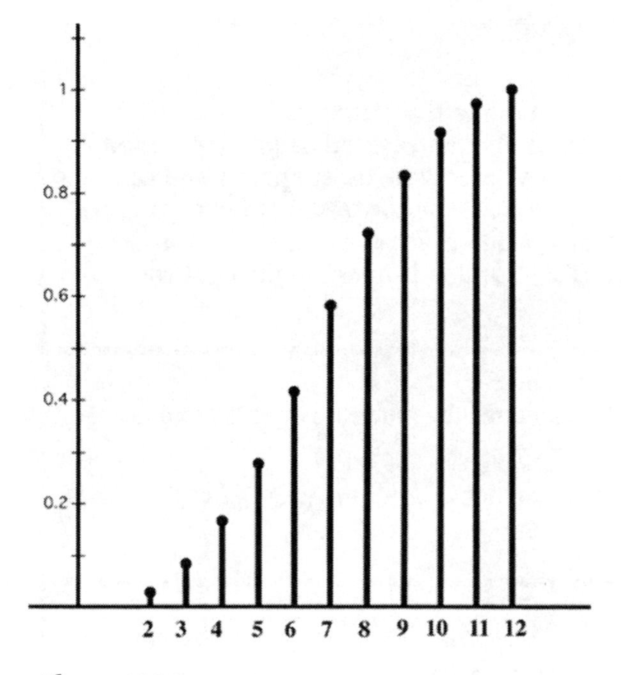

Figure 7.4
The cumulative distribution function of the random variable *X* that gives the sum of two dice.

- The value of $F(a)$ is between 0 and 1 because it is a probability. This means for any number a,

$$0 \leq F(a) \leq 1.$$

- If you know the probability density function, you can compute the cumulative distribution function. To get $F(a)$ for any number a, just add up the probabilities that the random variable has values up to and including a. This is just the sum

$$F(a) = \sum_{x \leq a} P\{X = a\}.$$

- You can find the probability that the random variable X takes on a value in a half-open interval $(a, b]$ from the cumulative distribution function. The value on the half-open interval is just a difference

$$\mathrm{P}\{a < X \leq b\} = F(b) - F(a)$$

> **NOTE**
>
> Recall that a **closed interval** of the real numbers $\{x \mid a \leq x \leq b\}$ includes its endpoints and can be represented as $[a, b]$. An **open interval** $\{x \mid a < x < b\}$ does not include its endpoints and can be represented as (a, b). The **half-open interval** $\{x \mid a < x \leq b\}$ that is open on the left can be represented as $(a, b]$ and the half-open interval $\{x \mid a \leq x < b\}$ that is open on the right can be represented as $[a, b)$.

- You can find a particular value for the random variable from the cumulative distribution function.

$$P\{X = a\} = P\{a - 1 < X \leq a\} = F(a) - F(a - 1)$$

> **NOTE**
>
> Each random variable has an associated probability frequency function and an associated cumulative distribution function. To apply a random variable to a specific problem, you must have the probability frequency function or the cumulative distribution function. Because of this close connection and because the distribution function is so important in applications, the term "distribution" is used when referring to the distribution *or* to the random variable. This does not result in any confusion because a random variable is very different from its distribution—a random variable is a function that assigns a number to each element of the sample space, and a distribution gives the probability that the random variable it is associated with will take on values in a given interval.

7.4 Joint Distribution of Two Random Variables

We have already seen that we can have two different random variables associated with the same sample space. In the examples given in Section 7.2, we have the random variable X that gives the sum of two dice and the random variable Y that gives the lower number shown on the dice.

When there are two random variables on the same sample space, we study their interaction using a **joint distribution**. For example, we might want to know the probability that the sum of the dice is 6, so $X = 6$, *and* that the lowest die is 3, so $Y = 3$. We write this $P\{X = 6, Y = 3\}$.

The probability $P\{X = 6, Y = 3\}$ is 1 / 36 because, of the five outcomes that give a sum of 6, only one has the lower number equal to 3. Or if we count the outcomes with the lower number equal to 3 first, we get seven of them, namely (3, 3), (3, 4), (3, 5), (3, 6), (6, 3), (5, 3), and (4, 3). Only one has a sum of 6, and this again gives us the probability $P\{X = 6, Y = 3\} = 1 / 36$.

One way to show a joint distribution is in a table, like the one shown in Figure 7.5, where each entry is a probability. The entry in the first row and first column gives the probability $P\{X = 1, Y = 1\}$, the entry in the second row and first column gives the probability $P\{X = 2, Y = 1\}$, the entry in the first row and second column gives the probability $P\{X = 1, Y = 2\}$, and so on. We can summarize this by saying that the entry in the *i*th row and *j*th column gives the probability $P\{X = i, Y = j\}$.

X \ Y	1	2	3	4	5	6	row sum
2	1/36	0	0	0	0	0	1/36
3	2/36	0	0	0	0	0	2/36
4	2/36	1/36	0	0	0	0	3/36
5	2/36	2/36	0	0	0	0	4/36
6	2/36	2/36	1/36	0	0	0	5/36
7	2/36	2/36	2/36	0	0	0	6/36
8	0	2/36	2/36	1/36	0	0	5/36
9	0	0	2/36	2/36	0	0	4/36
10	0	0	0	2/36	1/36	0	3/36
11	0	0	0	0	2/36	0	2/36
12	0	0	0	0	0	1/36	1/36
column sum	11/36	9/36	7/36	5/36	3/36	1/36	

Figure 7.5
The joint distribution function of random variables *X* and *Y*.

For each row and column of the joint distribution table, the sums of all entries are given in the margin. The sum of all the entries in the first row gives the probability $P\{X = 1\}$, the sum of all the entries in the second row gives the probability $P\{X = 2\}$, and so on. Similarly, the sum of all the entries in the first column gives the probability $P\{Y = 1\}$, the sum of all the entries in the second column gives the probability $P\{Y = 2\}$, and so on. The sums that give the probabilities $P\{X = i\}$ or $P\{Y = j\}$ are called **marginal probabilities** because of their location.

You can check to see that in the joint distribution table the marginal probabilities for the columns or the rows add up to 1. This is because the marginal probabilities for the rows give the probabilities that Y will take on any of its values. The sum of the marginal probabilities for the columns gives the probability that X will take on any of its values. These sums are both 1; we can use this fact as a check that all entries in a joint distribution table have been entered correctly.

Also, the sum of all probabilities in the table is 1 because the entries give the probabilities for all possible events in the sample space.

Recall that two events A and B in the same sample space are independent if

$$P(AB) = P(A) \times P(B)$$

There is a similar concept for random variables. Two random variables X and Y are **independent** if

$$P\{X = a, Y = b\} = P\{X = a\} \times P\{Y = b\}$$

for all values of a and b.

For example, throw two dice. Consider the same two random variables, X, giving the sum of the two dice and Y giving the lower number on the two dice. It is easy to see that X and Y are not independent, since $P\{X = 2, Y = 3\} = 0$ but

$$P\{X = 2\} \times P\{Y = 3\} = (1 / 36) \times (7 / 36) = 7 / 1296$$

On the other hand, suppose the dice are colored red and green. If W gives the value on the red die and Z gives the value on the green die, these random variables are independent. This is because $P\{W = a, Z = b\}$ is $1 / 36$ no matter what values a and b have, and this is the same as $P\{W = a\} \times P\{Z = b\} = (1 / 6) \times (1 / 6) = 1 / 36$.

7.5 The Bernoulli Random Variable and the Binomial Distribution

In many situations, there are only two outcomes to a trial or experiment, one with probability p and the other with probability $1 - p$. Often one outcome is called a **success** and the other a **failure**. When this is the case, p is used as the probability of success. For convenience, the probability of failure $1 - p$ is sometimes called q.

Such an experiment is called a **Bernoulli trial**, after the Swiss mathematician James Bernoulli, or a **binomial trial**, because there are only two outcomes.

The random variable associated with a Bernoulli trial is the Bernoulli random variable, with value 1 for a successful outcome and value 0 for failure.

Examples of Bernoulli Trials

- Flipping a coin gives two outcomes, heads and tails. If we assign the value 1 to heads and 0 to tails, we have a Bernoulli random variable. If we call this random variable X, then $P\{X = 1\} = 0.5$ and $P\{X = 0\} = 0.5$, since heads and tails are equally likely.
- Rolling a die has six outcomes, but if we decide to call the event $\{1, 2\}$ a success and call the event $\{3, 4, 5, 6\}$ a failure, then we have a Bernoulli trial. Let the associated random variable be Y. Then $P\{Y = 1\} = 1 / 3$ and $P\{Y = 0\} = 2 / 3$.

If we repeat a Bernoulli trial many times over, we get a new situation, called a **binomial experiment**. The number of trials n and the probability p are its **parameters**, numbers that define the structure or properties of the experiment. If there are n trials and each trial has probability of success equal to p and failure equal to $1 - p$, the binomial experiment is denoted $B(n, p)$. Counting the number of successes in a binomial experiment gives us a new random variable, the **binomial random variable** with parameters (n, p).

For example, if we flip a coin eight times and heads is considered a success, we have the binomial experiment B(8, 0.5), and the associated random variable that gives the number of heads in the eight flips has parameters (8, 0.5).

To compute the probability of k successes in n trials, first count up how many ways you can choose the k successes and the $n - k$ failures. This is just the number of ways to choose k objects from a collection of n objects, ${}_nC_k$. Each of these choices of k objects gives an event with k successes and $n - k$ failures. Because success has probability p and there are k successes, the probability of k successes is p^k.

Similarly, the probability of the $n - k$ failures for each of these choices is $(1 - p)^{(n-k)}$. Thus for the binomial random variable X that gives the number of successes for n trials,

$$P\{X = k\} = {}_nC_k\, p^k\, (1 - p)^{(n-k)}$$

Now you can see why this is called a *binomial* random variable—computation of the probabilities uses the binomial coefficients ${}_nC_k$.

For the binomial experiment B(8, 0.5), we have, for example,

$$P\{X = 3\} = {}_8C_3\, (0.5)^3\, (0.5)^5 = 56 \times 0.125 \times 0.03125 = 0.21875.$$

Examples of the Binomial Distribution

- What is the probability of getting exactly 10 heads when you flip a coin 20 times? Of getting exactly eight heads?

  Solution: Let getting a head be a success; the probability of a head is $p = 0.5$.

  $$P\{X = 10\} = {}_nC_k\, p^k\, (1 - p)^{(n-k)} = {}_{20}C_{10}\, (0.5)^{10}\, (0.5)^{10}$$
  $$\approx 184{,}756 \times 0.0009766 \times 0.0009766 \approx 0.176$$
  $$P\{X = 8\} = {}_nC_k\, p^k\, (1 - p)^{(n-k)} = {}_{20}C_8\, (0.5)^8\, (0.5)^{12}$$
  $$\approx 125{,}970 \times 0.00391 \times 0.000244 \approx 0.120$$

- Roll one die eight times. What is the probability of getting 1 or 2 on three of the rolls?

  Solution: Let getting a 1 or 2 be success. The probability of success is $2 / 6 = 1 / 3 \approx 0.333$.

  $$P\{Y\} = {}_nC_k\, p^k\, (1 - p)^{(n - k)} = {}_8C_3\, (1 / 3)^3\, (2 / 3)^5$$
  $$\approx 56 \times 0.037 \times 0.132 \approx 0.273$$

We can use the binomial distribution only when there are repeated, independent trials, each having the same probability of success. The following are examples when the binomial distribution is appropriate:

- Flipping a coin
- Tossing dice
- Playing a game of chance where the probability of winning (success) is the same in each game
- Factory production where the probability of producing a defective item ("success") stays the same
- Data transmission where the chance of an error ("success") is the same for each bit

The following are examples when it might *not* be appropriate to use the binomial distribution:

- The probability of rain on a given day. (The probabilities are different on different days and the trials are not independent, as it is less likely to rain after a sunny day than after a rainy day.)
- Errors made by a student on a long multiple-choice exam. (If the questions get harder or if the student gets tired, the probability of success on later questions will be less than on earlier questions.)
- Winning a sports game where conditions, such as the starting lineup, weather, location, and so on, change from game to game

7.6 The Poisson Distribution

People call into a help desk at random. A marine biologist counts the number of certain microorganisms in a sample of pond water. A new flash drive is defective. A physicist counts the number of photons hitting a detector every minute. A quality inspector finds flaws in a piece of fabric.

These are quite different from most of the examples that we have seen so far. There is no sample space. Rather, there is a time interval or region in space that we are looking at, and specific outcomes happen, seemingly at random. For these kinds of situations, we use the Poisson random variable, named after the French mathematician Siméon-Denis Poisson.

The Poisson random variable is used when the following criteria are met:

- There is a relatively small number of successes or occurrences in a large interval of time or in a large region of space.
- The number of occurrences in two intervals or regions are independent of each other as long as the intervals or regions do not overlap.
- The average number of occurrences in intervals or regions of the same size is about the same throughout the experiment or observation.
- Two occurrences do not happen simultaneously.

The value of the Poisson random variable X can be any non-negative integer, 0, 1, 2, 3, 4, and so on. The formula that gives the probability distribution depends on the average rate of occurrences, given as outcomes or successes per time or region; this rate is usually small and may not be an integer. In a specific situation, this rate is determined empirically from observations or measurements. We use the Greek letter lambda, λ, for the average rate of the outcomes that we are counting. In applications, this number must be determined empirically, from observation.

For average rate λ and interval of size t, the Poisson distribution gives the probability of k successes over a region or interval of size t as

$$P\{X = k\} = e^{-\lambda t}\frac{(\lambda t)^k}{k!}$$

Poisson derived this formula using techniques of calculus. The letter e in this formula is the number 2.71828. . . , which is the base for natural logarithms. Most calculators have a function that gives e^x.

To see how to use the Poisson distribution, suppose that we know there are 400 misprints in a book that is 600 pages long. The misprints are the "successes," or occurrences, that we are counting and the region is a page of the book. The average rate of misprints is determined from the given data, and is $\lambda = 400 / 600 \approx 0.667$ misprints per page. The Poisson distribution is appropriate here if we assume that the misprints are randomly distributed.

For the formula, we have $\lambda = 0.667$ and $t = 1$ page, so

$$P\{X = k\} = e^{-0.667} \frac{(0.667)^k}{k!}$$

We get the following values for $k = 1, 2, 3, 4, 5$, and 6:

$$P\{X=0\} = \frac{e^{-0.667}(0.667)^0}{0!} = \frac{e^{-0.667} \times 1}{1} \approx 0.513$$

$$P\{X=1\} = \frac{e^{-0.667}(0.667)^1}{1!} = \frac{e^{-0.667} \times (0.667)}{1} \approx 0.342$$

$$P\{X=2\} = \frac{e^{-0.667}(0.667)^2}{2!} \approx 0.114$$

$$P\{X=3\} = \frac{e^{-0.667}(0.667)^3}{3!} \approx 0.0254$$

$$P\{X=4\} = \frac{e^{-0.667}(0.667)^4}{4!} \approx 0.00423$$

$$P\{X=5\} = \frac{e^{-0.667}(0.667)^5}{5!} \approx 0.000565$$

$$P\{X=6\} = \frac{e^{-0.667}(0.667)^6}{6!} \approx 0.0000628$$

From this data we see that it is most likely there is no misprint on a page; this happens with probability 0.513. The probability of one misprint is 0.342, and as the number of misprints increases, the probability decreases dramatically. In fact, even for six misprints, the probability is so small (less than 1 / 10,000) that we don't expect to find six misprints on one page, and it is even less likely to find more than six misprints on a page.

Examples where you could use the Poisson distribution:

- The number of red candies in a bag of multicolored candies
- The number of houses sold by a realtor in a given month
- The number of customers coming into a store in an hour
- The number of defective items in a sample taken from the day's production
- The number of genetic mutations per generation in an experiment with fruit flies

Some examples of when *not* to use the Poisson distribution:

- The number of hits per day on a new website. (The number is not independent from day to day; for a good website, it should go up every day.)
- The number of hits per hour on a successful website. (The number of hits would be very large and could vary depending on the time of day. Also, hits could be simultaneous.)
- The number of orders placed on a website per hour. (The number could depend very much on the time of day or day of the year.)
- The weight of defective items from a day's production. (This is not an integer.)
- The number of students each day who are at home with the flu. (The number is not independent from day to day, as the flu is contagious.)

Because the Poisson distribution is used for relatively rare occurrences, it is sometimes called the **Law of Small Numbers**.

Up until the Poisson distribution, all the probabilities we have studied have been computed using the number of successes divided by the total number of outcomes; these probabilities are intuitively clear. The probabilities in the Poisson model are quite different, being given by a formula. However, the Poisson distribution has been repeatedly shown to effectively model real-life situations that have the criteria previously listed, and it is frequently used by scientists and engineers.

Examples of the Poisson Distribution

- Left-handed people make up about 10% of the population. A school is buying arm-desks for a new classroom. The classroom has 30 desks, and the school wants to put two left-handed desks and 28 right-handed desks in the classroom. What is the probability that there will be more than two left-handed students in the next class of 30?

  The rate of left-handed people is $\lambda = 0.10$, and the value of t is 30 because there are 30 students in the class; thus, $\lambda t = 3$. The probability that k students are left-handed is

$$P\{X = k\} = e^{-3}\frac{3^k}{k!}$$

  Find the complement of the event that 0, 1, or 2 students will be left-handed. This is

$$\begin{aligned} 1 - \left(\mathrm{P}\{X = 0\} + (\mathrm{P}\{X = 1\} + (\mathrm{P}\{X = 2\}\right) &= 1 - \left(e^{-3}\frac{3^0}{0!} + e^{-3}\frac{3^1}{1!} + e^{-3}\frac{3^2}{2!}\right) \\ &\approx 1 - \left(0.0498 + 0.149 + 0.224\right) \\ &\approx 0.577 \end{aligned}$$

  This is a fairly large probability, so the school should consider buying more than two left-handed desks for the classroom.

- After counting many bags of candy, Jella has found that there is an average of 10 red candies per bag of 100 candies. Her sister has eaten half of the bag that Jella had in her room. There are three red candies left in the bag. Is Jella justified in claiming that her sister took more than her share of red candies?

  Let the random variable R give the number of red candies. The rate of red candies is $\lambda = 10 / 100 = 0.1$, and the value of t is 50 because there is half of the bag—50 candies—left. The value of λt is $0.1 \times 50 = 5$, so

  $$P\{R = 3\} = e^{-5} \times (5^3 / 3!) \approx 0.140.$$

  This is a small number, so Jella feels justified in thinking her sister was picking out the red candies.

  However, Jella's sister feels that Jella's conclusion is not valid, and she decides to compute the probability that there would be more than three left. It is easiest to use the complement of that event to find the probability that there are 0, 1, 2, or 3 left, and then subtracting from 1. This is

  $$\begin{aligned} &1 - \left(P\{X=0\} + P\{X=1\} + P\{X=2\} + P\{X=3\}\right) \\ &= 1 - \left(e^{-5}\frac{5^0}{0!} + e^{-5}\frac{5^1}{1!} + e^{-5}\frac{5^2}{2!} + e^{-5}\frac{5^3}{3!}\right) \\ &\approx 1 - \left(0.00674 + 0.0337 + 0.0842 + 0.140\right) \\ &\approx 1 - 0.265 \\ &\approx 0.735 \end{aligned}$$

  This means the probability that there should be more than three red candies left is 0.735. Jella's sister decides it is better not to say anything.

7.7 The Geometric Distribution

Suppose we flip a coin until we get a head for the first time. How many flips will it take? With probability 0.5, we will get a head on the first flip. With probability $0.5 \times 0.5 = 0.25$, we get tails then heads, thus getting the first head on the second flip. The geometric distribution describes this kind of situation, when we repeat binomial trials until we get a success.

If the probability of success in a binomial trial is p, then the **geometric random variable** gives the number of trials until the first success. The probability of $n - 1$ failures followed by a success on the nth trial is $P\{X = n\}$. A geometric random variable is sometimes called a **Pascal random variable**.

For example, flip a coin until you get a head. The probability of success, of getting a head, is 0.5. Let the random variable X give the number of trials it takes to get the first success. Then $X\{H\} = 1$ and $P\{X = 1\} = 0.5$, $X\{TH\} = 2$ and $P\{X = 2\} = 0.25$, $X\{TTH\} = 3$ and $P\{X = 3\} = 0.125$. The probability of getting the first head on the fifth flip is $(0.5)^4 \times (0.5) = 0.03125$ because the probability of failure on the first four flips is $(0.5)^4$, and the probability of success on the last flip is 0.5. In symbols, this is $P\{X = 5\} = (0.5)^4 \times (0.5)$.

This reasoning gives us the general formula for the probability of the first success on the nth trial:

$$P\{X = n\} = (1 - p)^{(n-1)} \times p$$

Example of the Geometric Distribution

Roll a die until you get a 6. What is the probability that this will happen on the sixth roll? What is the probability that it will happen before the sixth roll?

The probability of getting a 6 is $p = 1/6 \approx 0.1667$ and $1 - p = 5/6 \approx 0.8333$. The probability that the first 6 will be on the sixth roll is $P\{X = 6\} \approx (0.8333)^5 \times 0.1667 \approx 0.06698$. The probability that it will happen *before* the sixth roll is

$$\begin{aligned} & P\{X=1\} + P\{X=2\} + P\{X=3\} + P\{X=4\} + P\{X=5\} \\ &\approx 0.1667 + \big((1 - 0.1667)\times 0.1667\big) + \big((0.8333)^2 \times 0.1667\big) \\ &\quad + \big((0.8333)^3 \times 0.1667\big) + \big((0.8333)^4 \times 0.1667\big) \\ &\approx 0.1667 + 0.1389 + 0.1157 + 0.0965 + 0.0804 \\ &\approx 0.5982 \end{aligned}$$

7.8 The Negative Binomial Distribution

We use the geometric distribution if we repeat binomial trials until we get the first success. What if we want to repeat the binomial trial until we get two success? Or five successes?

For example, flip a coin until you get five heads. The probability of success, of getting a head, is 0.5. To compute the probability of getting five heads in 10 flips, we first choose which of the 10 flips gives heads. One of the heads, of course, must happen on the last flip. There are ${}_9C_4 = 126$ ways to choose four successes in the first nine flips. The probability of getting five successes in 10 flips is then $126 \times (0.5)^4 \times (0.5)^5 \times 0.5 = 0.123$.

The negative binomial distribution is similar to the geometric distribution, except now we are interested in repeating the binomial trials until we get r successes. Suppose the probability of success in a binomial trial is p. The **negative binomial random variable** gives the number of trials until we achieve r successes. So if the value of the random variable is n, that means there are n trials, r successes, and one of the successes happens on the nth or last trial.

For example, roll a die until 1 comes up three times. What is the probability that this will take five rolls? The probability of success, of getting a 1, is $1 / 6 \approx 0.1667$. The probability of failure, getting anything but 1, is $5 / 6 \approx 0.8333$. The probability of getting a 1 on the fifth roll is 0.1667. To find the probability on getting two 1s in the first four rolls, we first count up how many ways this can be done, ${}_4C_2 = 6$. Then we find the probability of two successes and two failures, $(0.1667)^2 \times (0.8333)^2 = 0.01929$. We multiply these three numbers to get the probability of three 1s in exactly five rolls, $0.8333 \times 6 \times 0.01929 \approx 0.9645$.

This reasoning gives us the general formula for the probability of getting r successes in exactly n trials, which is

$$P\{X = n\} = {}_{n-1}C_{r-1} \times p^r \times (1 - p)^{(n-r)}$$

Example of the Negative Binomial Distribution

Roll a die until you get four 6s. What is the probability that this will happen in seven rolls? What is the probability that it will happen in 10 rolls?

The probability of getting a 6 is $p = 1/6 \approx 0.1667$. The probability of getting four 6s in the first seven rolls can be computed from the formula with $n = 7$ and $r = 4$. The probability is

$$\begin{aligned} P\{X = 7\} &\approx {}_6C_3 \times 0.1667^4 \times (0.8333)^3 \\ &\approx 20 \times 0.000772 \times 0.579 \approx 0.00893 \end{aligned}$$

For the probability of four 6s in 10 rolls, n is 10, but r is still 4. The probability is

$$\begin{aligned} P\{X = 10\} &\approx {}_9C_3 \times 0.1667^4 \times (0.8333)^6 \\ &\approx 84 \times 0.000772 \times 0.335 \approx 0.0217 \end{aligned}$$

7.9 The Hypergeometric Distribution

Suppose that you choose four marbles from an urn containing 20 marbles, six white and 14 red. You want to get three white and one red. What are your chances? We compute this as the ratio of successes to the total number of outcomes. The number of successes is the number of ways of choosing three white marbles times the number of ways of choosing one white marble, or ${}_6C_3 \times {}_{14}C_1$. The number of possible outcomes is the number of ways of choosing four marbles from twenty, or ${}_{20}C_4$. The probability is

$$\frac{{}_6C_3 \times {}_{14}C_1}{{}_{20}C_4} = 0.0578$$

The **hypergeometric random variable** gives the number of successes when making n choices from a collection of N objects, with probability of success equal to p. There are Np objects that are successes, and if you want k of those objects, there are ${}_{Np}C_k$ ways to get them. There are $N - Np$ other objects, and ${}_{N-Np}C_{N-k}$ ways to choose those. There are ${}_NC_n$ ways to choose n objects from the N objects, and this is the number of outcomes in the sample space.

This reasoning leads to the hypergeometric distribution:

$$P\{X = k\} = \frac{{}_{Np}C_k \times {}_{N-Np}C_{n-k}}{{}_{N}C_n}$$

This is often referred to as **sampling without replacement**, since you could have chosen the marbles one by one from the urn without putting any back in. Compared to this, binomial trials are referred to as sampling *with* replacement, since each trial has the same probability of success.

Examples of the Hypergeometric Distribution

- You are dealt six cards from a standard deck. Let X be the random variable that gives the number of spades. The probability of getting exactly four spades is $P\{X = 4\} = {}_{13}C_4 \times {}_{39}C_2 / {}_{52}C_6 = 715 \times 741 / 20{,}358{,}520 \approx 0.0260$.
- A shipment of computer chips from a supplier is expected to contain about 2% defective chips. From a lot of 500 chips, three are taken out. What is the probability that all three will be not defective?

  Let D be the random variable that gives the number of defective chips. The number of objects N in this case is 500 and p is 0.02, so $Np = 10$. We want $P\{D = 0\} = {}_{10}C_0 \times {}_{490}C_3 / {}_{500}C_3 = 1 \times 19{,}488{,}280 / 20{,}708{,}500 \approx 0.941$.

7.10 The Multinomial Distribution

The binomial distribution or the hypergeometric distribution is used when there are two outcomes, success and failure. We have seen cases, such as throwing a die, when there are more than two outcomes. If we throw a die many times in succession, we may want to know the probability of a particular outcome. For example, if we throw a die 10 times, what is the probability of getting a 2 three times, a 3 three times, and a 6 four times? We know there are 6^{10} outcomes. To count up the number of successes, we use the Partition Rule, discussed in Chapter 4. We count the number of ways to partition the 10 outcomes into sets of 3, 3, and 4.

This is 10! / (3! 3! 4!) = 4,200. The probability of getting 2 three times, a 3 three times, and a 6 four times is therefore 4,200 / $6^{10} \approx 0.0000695$.

In the more general case, each of the different outcomes could have different possibilities. Suppose there is an experiment that has r different outcomes with different probabilities $p_1, p_2, p_3, \ldots, p_r$ for each outcome. Let the random variable X_i give the number of times the ith outcome happens. To compute the probability

$$P\{X_1 = n_1, X_2 = n_2, X_3 = n_3, \ldots, X_r = n_r\}$$

we first need to count how many ways we can partition the $n_1 + n_2 + n_3 + \cdots + n_r$ outcomes into sets of $n_1, n_2, n_3, \ldots,$ and n_r outcomes. By the Partition Rule, this is

$$\frac{n!}{n_1!n_2!n_3!\ldots n_r!}$$

The probability of getting the first outcome n_1 times is $p_1^{n_1}$, the probability of getting the second outcome is $p_2^{n_2}$, and so on. Finally, we get the probability

$$P\{X_1 = n_1, X_2 = n_2, X_3 = n_3, \ldots, X_r = n_r\} = \frac{n!}{n_1!n_2!n_3!\ldots n_r!} p_1^{n_1} p_2^{n_2} p_3^{n_3} \ldots p_r^{n_r}$$

This is the **multinomial random distribution**, the joint distribution of r different random variables.

Example of the Multinomial Distribution

Marlee chooses 10 candies from a bag of 100 candies. There are 10 red, 20 yellow, 20 green, 25 orange, and 25 pink candies. What is the chance that she gets 3 red, 3 yellow, and 4 orange candies?

The probability of a red candy is 10 / 100 = 0.10; of a yellow candy, 20 / 100 = 0.20; of a green candy, 0.20; of an orange candy, 0.25; and of a pink candy, 0.25. The chance of getting a 3 red, 3 yellow, and 4 orange candies is

$$\frac{10!}{3!3!4!}(0.10)^3(0.20)^3(0.25)^4 = 0.0007875$$

Chapter 7 Summary

A **random variable** is a rule or function that assigns a number to a random outcome. Random variables are also called **stochastic variables** or **variates**.

Discrete random variables take on only a few values or only integer values.

The set of different values that a random variable can take on is called the **range** or the **range space** of the random variable.

The notation $P\{X = 6\} \approx 0.139$ means that the probability that the random variable X takes on value 6 is about 0.139

The function that assigns a probability to each number in the range of a random variable is called the **probability frequency function**, the **probability function**, or **probability mass function**.

Important properties of random variables:

- All probabilities have values in the interval from 0 to 1.

$$0 \leq P\{X = a\} \leq 1$$

- The sum of the probabilities for all outcomes in any sample space is 1.

$$\sum_a P\{X = a\} = 1$$

- The outcomes $X \leq a$ and $X > a$ are complementary events for any number a.

$$P\{X \leq a\} = 1 - P\{X > a\}$$

For a given value x of the random variable, the **cumulative distribution function** gives the probability of the random variable taking on a value up to and including the given value x, $F(x) = P\{X \leq x\}$.

The cumulative distribution function is also called the **CDF**, **probability distribution**, or **distribution function**.

Important properties of cumulative distribution functions:

- The cumulative distribution function $F(x)$ is a **nondecreasing** function. For numbers a and b,

$$F(a) \leq F(b) \text{ if } a < b$$

- The value of $F(x)$ is between 0 and 1.

$$0 \leq F(x) \leq 1 \text{ for all numbers } x$$

- The cumulative distribution function can be computed from the probability density function.

$$F(a) = \sum_{x \leq a} P\{X = a\}$$

- The probability that the random variable X takes on a value in the half-open interval $(a, b]$ can be computed from the cumulative distribution function.

$$P\{a < X \leq b\} = F(b) - F(a)$$

- A particular value for the random variable can be determined from the cumulative distribution function.

$$P\{X = a\} = P\{a - 1 < X \leq a\} = F(a) - F(a - 1)$$

Two random variables on the same sample space have a **joint distribution**. The notation $P\{X = 6, Y = 3\} \approx 0.278$ means that the probability that the random variable X takes on value 6 and the random variable Y takes on value 3 is about 0.278.

The values of a joint distribution can be displayed in a table where each entry is a probability. The probability $P\{X = i, Y = j\}$ is the entry in the ith row and jth column.

For each row and column of the joint distribution table, the sums of all entries are given in the margin. These sums give the probabilities $P\{X = i\}$ or $P\{Y = j\}$ and are called **marginal probabilities** because of their location.

Two random variables X and Y are **independent** if

$$P\{X = a, Y = b\} = P\{X = a\} \times P\{Y = b\}$$

for all values of a and b.

If there are only two outcomes to a trial or experiment, one with probability p is called a **success** and the other with probability $1 - p$ is called a **failure**. Such an experiment is called a **Bernoulli trial** or a **binomial trial**.

The random variable associated with a Bernoulli trial is the **Bernoulli random variable**, with value 1 for a successful outcome and value 0 for failure.

A Bernoulli trial repeated many times is called a **binomial experiment**. The number of trials n and the probability p are the **parameters** that define the experiment, which is denoted $B(n, p)$. The number of successes in a binomial experiment is the **binomial random variable** with parameters (n, p).

For the binomial random variable X that gives the number of successes for n trials,

$$P\{X = k\} = {}_nC_k\, p^k (1 - p)^{(n-k)}$$

The **Poisson distribution** is used for a random variable that meets the following criteria:

- There are relatively few successes or occurrences in a large interval of time or in a large region of space.
- The numbers of occurrences in two intervals or regions are independent if the intervals or regions containing them do not overlap.
- The average number of occurrences in intervals or regions of the same size is about the same throughout the experiment or observation.
- Two occurrences do not happen simultaneously.

If a Poisson distribution has average rate of successes λ and interval of size t, then

$$P\{X = k\} = e^{-\lambda t} \frac{(\lambda t)^k}{k!}$$

The Poisson distribution is sometimes called the **Law of Small Numbers**.

If the probability of success in a binomial trial is p, then the **geometric random variable** gives the number of trials until the first success. It is computed using the formula

$$P\{X = n\} = (1 - p)^{(n-1)} \times p$$

If the probability of success in a binomial trial is p, then the **negative binomial random variable** gives the number of trials until r successes. It is computed using the formula

$$P\{X = n\} = {}_{n-1}C_{r-1} \times p^r \times (1 - p)^{(n-r)}$$

The **hypergeometric random variable** or **sampling without replacement** gives the number of successes k when making n choices from a collection of N objects with probability of success equal to p. It is computed using the formula

$$P\{X = k\} = \frac{{}_{Np}C_k \times {}_{N-Np}C_{n-k}}{{}_{N}C_n}$$

The **multinomial random distribution** gives the joint distribution of r different random variables. If there are r different outcomes with different probabilities $p_1, p_2, p_3, \ldots, p_r$ for each outcome and the random variable X_i gives the number of times the ith outcome happens, then

$$P\{X_1 = n_1, X_2 = n_2, X_3 = n_3, \ldots, X_r = n_r\} = \frac{n!}{n_1!n_2!n_3!\ldots n_r!} p_1^{n_1} p_2^{n_2} p_3^{n_3} \ldots p_r^{n_r}$$

Chapter 7 Practice Problems

1. Toss three coins. The random variable Y gives the number of tails.
   a. Find Y(HTT)
   b. Find Y(THT)
   c. What is the range of Y?
   d. Find $P\{Y = 0\}$
   e. Find $P\{Y = 1\}$
   f. Find $P\{Y = 2\}$
   g. Find $P\{Y = 3\}$
   h. Find $P\{Y \geq 2\}$
   i. Find $P\{1 \leq Y \leq 2\}$
   j. Graph the probability distribution for the random variable Y.
   k. What information can we get about the random variable Y from the graph?
2. Draw a card from a standard deck. The random variable Z gives the number on the card or gives 10 if the card is a face card.
   a. Find $Z(3\blacklozenge)$
   b. Find $Z(8\blacklozenge)$
   c. Find $Z(K\blacklozenge)$

d. What is the range of Z?

e. Find $P\{Z = 1\}$

f. Find $P\{Z = 5\}$

g. Find $P\{Z = 10\}$

h. Find $P\{6 \leq Z\}$

i. Find $P\{3 \leq Z \leq 8\}$

j. Graph the probability distribution for the random variable Z.

k. What information can we get about the random variable Z from the graph?

3. Throw three dice. Compute the values for the random variable W that gives the sum of three dice for $P\{W = n\}$ for n from 7 to 15. (This finds the remaining values for the example in Section 7.2)

4. Answer the following questions for the joint distribution of two random variables X and Y shown in Figure 7.6.

a. What information does the number 0.2 in the table give?

b. What is $P\{X = 1, Y = 3\}$?

c. Which outcome(s) is most likely?

d. Which outcome(s) is least likely?

e. What information does the number 0.4 in the table give?

f. What information does the number 0.6 in the table give?

X \ Y	0	1	2	3	row sum
1	0	0.1	0	0.05	0.15
2	0.15	0.1	0	0	0.25
3	0.1	0.2	0.3	0	0.6
column sum	0.25	0.4	0.3	0.05	

Figure 7.6
The joint distribution of two random variables X and Y.

5. The table in Figure 7.7 shows some of the values for the joint distribution of two random variables X and Y. You can use this information along with what you know about probabilities and joint distributions to determine the missing values and answer the following questions.
   a. What is $P\{X = 2, Y = 3\}$?
   b. What is $P\{X = 3, Y = 4\}$?
   c. Fill in the rest of the joint distribution table.
   d. What is $P\{X = 1\}$?
   e. What is $P\{Y = 1\}$?
   f. Which X and Y pair(s) are most likely?
   g. Which X and Y pairs(s) are least likely?

Y \ X	1	2	3	4	row sum
0			0.1	0.1	
1	0.2	0		0.05	0.35
2	0	0.05		0.1	0.2
3		0	0		0.1
column sum	0.35			0.3	

Figure 7.7
The joint distribution function of random variables X and Y.

6. Throw a die and let 1 be a success and let 2, 3, 4, 5, and 6 be failures. This is a Bernoulli trial.
   a. What is the probability of success?
   b. What is the probability of failure?
   c. What does B(6, 0.1667) mean in this context?
   d. For X associated with B(6, 0.1667), compute $P\{X = 4\}$.
   e. For Y associated with B(4, 0.1667), compute $P\{Y = 1\}$.
7. A message 10 bits long is sent. The probability of error is 0.002 for each bit. What is the probability that the message will be sent with at most one error?

8. A random variable Y has a Poisson distribution $P\{Y = k\} = e^{-0.3} (0.3)^k / k!$. Find $P\{Y = 3\}$.
9. It is known that there are 1,000 fish in a pond and 20 have been tagged. You have caught three fish. What is the probability that at least one of them is tagged?
10. A county has an average rate of three thunderstorms per week during the summer. What is the probability that there will be at least three thunderstorms during the next two weeks?
11. Flip a coin until you get a head. What is the probability that this will happen on the first flip? Second flip? Third flip?
12. Toss two dice until you get 3 or 6. What is the probability that this will happen on the first toss? Second toss? Third toss?
13. Flip a coin until you get seven heads. What is the probability that this will happen with 10 flips? Fifteen flips? Twenty flips?
14. Toss two dice until you get a 3 or 6 four times. What is the probability that this will happen on four tosses? Eight tosses? Ten tosses?
15. A bag of candies has 15 red, five yellow, 10 blue, and 20 green. Choose five candies at random. What is the probability that you will get one red, one yellow, one blue, and two green?
16. Toss a die six times. What is the probability of getting three 2s and three 3s?
17. For each of the following situations, find at least one quantity that you could measure that would be a random variable. For most of these, there are many different possibilities.
    - **a.** Roll two dice
    - **b.** Choose a student at random from a class
    - **c.** Take five marbles from an urn that has red, blue, white, and green marbles
    - **d.** Take a sample of water from a pond
    - **e.** Take a sample of earth from a garden
    - **f.** Look at the purchases at a department store on a day chosen at random
    - **g.** Choose a batch of 10 keyboards from an assembly line that produces hundreds of keyboards a day
    - **h.** Choose a family at random from a city

Answers to Chapter 7 Practice Problems

1. **a.** $Y(\text{HTT}) = 2$
   b. $Y(\text{THT}) = 2$
   c. The range of Y is $\{0, 1, 2, 3\}$.
   d. $P\{Y = 0\} = 1 / 8 = 0.125$
   e. $P\{Y = 1\} = 3 / 8 = 0.375$
   f. $P\{Y = 2\} = 3 / 8 = 0.375$
   g. $P\{Y = 3\} = 1 / 8 = 0.125$
   h. $P\{Y \geq 2\} = 3 / 8 + 1 / 8 = 4 / 8 = 0.5$
   i. $P\{1 \leq Y \leq 2\} = 3 / 8 + 3 / 8 = 6 / 8 = 0.75$
   j. See Figure 7.8.
   k. The graph is symmetric, so $P\{Y = 0\} = P\{Y = 3\}$ and $P\{Y = 1\} = P\{Y = 2\}$. The probability is greatest for one or two heads and least for zero or three heads.

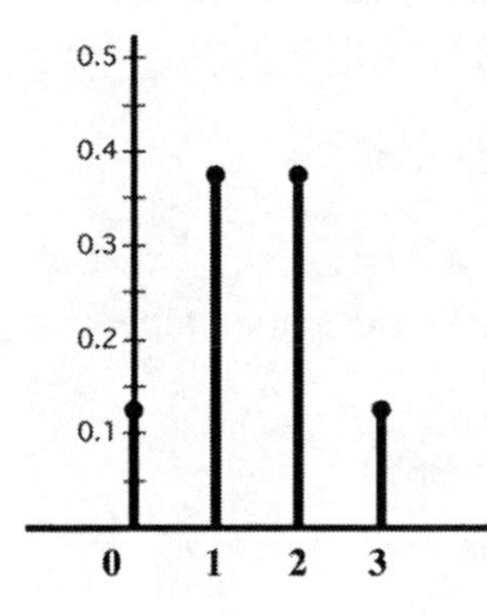

Figure 7.8
The probability distribution of the random variable *Y*.

2. **a.** $Z(3\blacklozenge\} = 3$
   b. $Z(8\blacklozenge\} = 8$
   c. $Z(\text{K}\blacklozenge\} = 10$
   d. The range of Z is $\{1, 2, 3, 4, 5, 6, 7, 8, 9, 10\}$.
   e. $P\{Z = 1\} = 4 / 52 \approx 0.0769$
   f. $P\{Z = 5\} = 4 / 52 \approx 0.0769$
   g. $P\{Z = 10\} = 12 / 52 \approx 0.231$

h. $P\{8 \leq Z\} = 32 / 52 \approx 0.615$

i. $P\{3 \leq Z \leq 8\} = 24 / 52 \approx 0.462$

j. See Figure 7.9.

k. All values for the random variable are the same except for $Z = 10$, where it is almost four times as large.

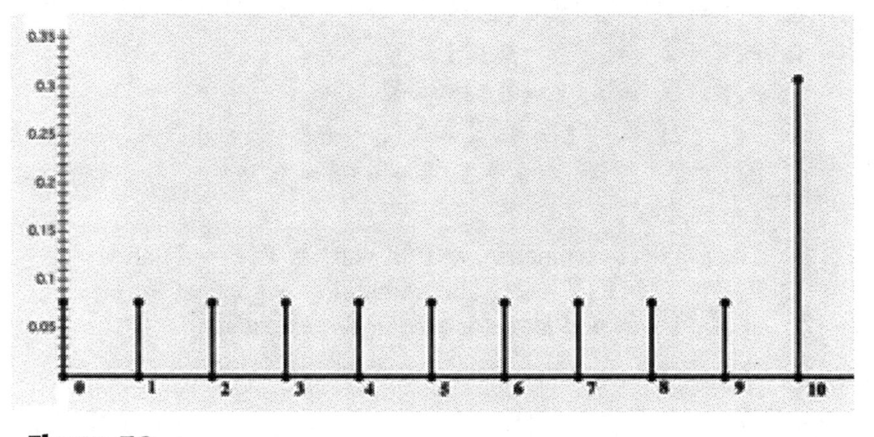

Figure 7.9
The probability distribution of the random variable *Z*.

3. Note that $P\{Y = n\}$ and $P\{Y = 21 - n\}$ have the same value; this reduces the number of computations needed.

$$P\{Y = 7\} = 15 / 216 \approx 0.0694$$
$$P\{Y = 8\} = 21 / 216 \approx 0.0972$$
$$P\{Y = 9\} = 25 / 216 \approx 0.116$$
$$P\{Y = 10\} = 27 / 216 = 0.125$$
$$P\{Y = 11\} = 27 / 216 = 0.125$$
$$P\{Y = 12\} = 25 / 216 \approx 0.116$$
$$P\{Y = 13\} = 21 / 216 \approx 0.0972$$
$$P\{Y = 14\} = 15 / 216 \approx 0.0694$$
$$P\{Y = 15\} = 10 / 216 \approx 0.0463$$

4. **a.** $P\{X = 3, Y = 1\} = 0.2$
   b. $P\{X = 1, Y = 3\} = 0.05$
   c. $X = 3$ and $Y = 2$
   d. There are five outcomes that will never happen: $X = 1$ and $Y = 0$; $X = 1$ and $Y = 2$; $X = 2$ and $Y = 2$; $X = 2$ and $Y = 3$; and $X = 3$ and $Y = 3$.
   e. $P\{Y = 1\} = 0.4$
   f. $P\{X = 3\} = 0.6$
5. **a.** $P\{X = 2, Y = 3\} = 0.05$
   b. $P\{X = 3, Y = 4\} = 0.05$
   c. See Figure 7.10. Use the fact that the sum of each row or column is a marginal probability and the sum of all the marginal probabilities for the rows or columns is 1.
   d. $P\{X = 1\} = 0.35$
   e. $P\{Y = 1\} = 0.35$
   f. The pair $X = 1$ and $Y = 1$ is most likely.
   g. The pairs $X = 1$ and $Y = 2$; $X = 2$ and $Y = 1$; $X = 2$ and $Y = 2$; and $X = 3$ and $Y = 3$ are least likely; in fact, these outcomes can never happen.

X \ Y	1	2	3	4	row sum
0	0.1	0.05	0.1	0.1	0.35
1	0.2	0	0.1	0.05	0.35
2	0	0.05	0.05	0.1	0.2
3	0.05	0	0	0.05	0.1
column sum	0.35	0.1	0.25	0.3	

Figure 7.10
The joint distribution of the random variables *X* and *Y*.

6. **a.** $1 / 6 \approx 0.1667$
   b. $5 / 6 \approx 0.8333$
   c. The binomial experiment consisting of 6 trials, with probability of success equal to 0.1667.
   d. $P\{X = 4\} = {}_6C_4\,(0.1667)^4\,(0.8333)^2 \approx 15 \times 0.0007722 \times 0.6944 \approx 0.008043$
   e. $P\{Y = 1\} = {}_4C_1\,(0.1667)^1\,(0.8333)^3 \approx 4 \times 0.1667 \times 0.5786 \approx 0.3858$
7. This is a binomial experiment consisting of 10 trials, with probability of "success" equal to 0.002. Call the random variable E. The probability of at most one error is

$$\begin{aligned} P\{E=0\} + P\{E=1\} &= {}_{10}C_0\,(0.002)^0\,(0.998)^{10} + {}_{10}C_1\,(0.002)^1\,(0.998)^9 \\ &= (1 \times 1 \times 0.9802) + (10 \times 0.002 \times 0.9821) \\ &\approx 0.9802 + 0.01964 \approx 0.9998 \end{aligned}$$

8. $P\{Y = 3\} = e^{-0.3}\,(0.3)^3 / 3! \approx 0.3334$
9. The rate of tagged fish is $20 / 1000 = 0.02$, so this is λ. Three fish have been caught, so $t = 3$ and $\lambda t = 0.06$, and the distribution is given by the formula

$$P\{X = k\} = e^{-0.06}(0.06)^k / k!$$

The probability that at least one has been tagged is

$$\begin{aligned} &P\{X=1\} + P\{X=2\} + P\{X=3\} \\ &\quad= e^{-0.06}(0.06)^1 / 1! + e^{-0.06}(0.06)^2 / 2! + e^{-0.06}(0.06)^3 / 3! \\ &\quad\approx 0.0565 + 0.00170 + 0.0000339 \\ &\quad\approx 0.0582 \end{aligned}$$

10. Let the random variable give the number of thunderstorms observed in two weeks. This has a Poisson distribution with $\lambda = 3$ and $t = 2$, so $\lambda t = 6$ and

$$P\{X = k\} = e^{-6}6^k / k!$$

The probability that there will be at least three thunderstorms during the next two weeks is

$$\begin{aligned} P\{X > 2\} &= 1 - P\{X=0\} - P\{X=1\} - P\{X=2\} \\ &= 1 - e^{-6}6^0 / 0! - e^{-6}6^1 / 1! - e^{-6}6^2 / 2! \\ &\approx 1 - 0.00248 - 0.0149 - 0.0446 \approx 0.938 \end{aligned}$$

11. Use the geometric distribution for the random variable X that gives the number of flips needed to get a head. In this example, $P\{X = n\} = (0.5)^{n-1} \times 0.5 = (0.5)^n$. $P\{X = 1\} = 0.5$, $P\{X = 2\} = 0.25$, and $P\{X = 3\} = 0.125$.

12. Use the geometric distribution for the random variable X that gives the number of tosses needed to get 3 or 6. The probability of *not* getting a 3 or 6 is $(4 \times 4) / 36 \approx 0.444$, and the probability of getting a 3 or 6 is about $1 - 0.444 \approx 0.556$. Thus $P\{X = n\} \approx (0.444)^{n-1} \times 0.556$. $P\{X = 1\} \approx 0.556$, $P\{X = 2\} \approx 0.444 \times 0.556 \approx 0.247$, and $P\{X = 3\} \approx (0.444)^2 \times 0.556 \approx 0.110$.

13. Use the negative binomial distribution for the random variable X that gives the number of flips needed to get seven heads.

$$P\{X = n\} = {}_{n-1}C^6 \times 0.57 \times 0.5^{(n-7)} = {}_{n-1}C_6 \times 0.5^n$$
$$P\{X = 10\} = {}_9C_6 \times 0.510 \approx 0.08203$$
$$P\{X = 15\} = {}_{14}C_6 \times 0.515 \approx 0.09164$$
$$P\{X = 20\} = {}_{19}C_6 \times 0.520 \approx 0.02588$$

14. Use the negative binomial distribution for the random variable X that gives the number of tosses needed to get 3 or 6 four times. The probability of success on one throw, as we saw in problem 12, is about 0.556 and the probability of failure is about 0.444.

$$P\{X = n\} \approx {}_{n-1}C_3 \times 0.556^4 \times 0.444^{(n-4)}$$
$$P\{X = 4\} \approx {}_3C_3 \times 0.556^4 \times 0.444^0 \approx 0.0424$$
$$P\{X = 8\} \approx {}_7C_3 \times 0.556^4 \times 0.444^4 \approx 0.130$$
$$P\{X = 10\} \approx {}_9C_3 \times 0.556^4 \times 0.444^6 \approx 0.0615$$

15. Use the multinomial distribution. Let random variables R, Y, B, and G be the number of candies of colors red, yellow, blue, and green, respectively, when five candies are chosen. Then

$$P\{R = 1, Y = 1, B = 1, G = 2\}$$
$$= \frac{5!}{1!1!1!2!}\left(\frac{15}{50}\right)^1\left(\frac{5}{50}\right)^1\left(\frac{10}{50}\right)^1\left(\frac{20}{50}\right)^2$$
$$\approx 0.0576$$

16. Use the multinomial distribution. There are six outcomes, each with probability 1 / 6. The answer is

$$\frac{6!}{1!3!3!1!1!1!}\left(\frac{1}{6}\right)^0\left(\frac{1}{6}\right)^3\left(\frac{1}{6}\right)^3\left(\frac{1}{6}\right)^0\left(\frac{1}{6}\right)^0\left(\frac{1}{6}\right)^0$$

$$= \frac{6!}{36} \times \frac{1}{216} \times \frac{1}{216}$$

$$\approx 0.0004287$$

17. These are only some of the examples of measurements that give random variables; other examples are also correct.

a. The sum of the dice; the larger of the two dice; the smaller of the two dice; the difference of the two dice

b. Height, weight, age, grade point average

c. The number colors in the sample; the number of red marbles; the number of marbles with the most frequent color

d. Number of some specific organism; pH; temperature; salinity

e. Salinity; organic matter content; pH; nitrogen content

f. Average receipt; total receipts for the day; number of customers

g. Number of defective keyboards

h. Income; number of family members; number of children; number of pets

Chapter

Continuous Random Variables

In Chapter 7, we looked at discrete random variables. In this chapter, we will look at random variables that can take on any value in an interval. For example, the weight of a battery or the length of its lifetime could be any number in an appropriately chosen interval.

A random variable like this, with values anywhere in an interval, is called a **continuous random variable**.

In this chapter, you will learn about continuous random variables and their properties and see some important examples of continuous distributions, including the uniform distribution, the normal distribution, and the exponential distribution.

8.1 Continuous Random Variables

If students in a probability class measure their heights, the measurements will probably land in the interval from 4 feet to 7 feet. The random variable that gives the height of a student chosen at random will have a value somewhere in the interval from 4 to 7. We use the notation [4, 7] to represent this interval. For the heights of students, we expect that most measurements will be clustered around an average value, and there will be fewer measurements as we go toward 4 feet or 7 feet.

If a highway tollbooth measures the amount of time between two cars, the values might range from 0 minutes to 60 minutes. The random variable that gives the time until the next car arrives will have a value in the interval [0, 60]. Knowledge of the properties of this random variable could help engineers determine whether to build another tollbooth.

If you choose a point at random on a line of length 1 meter, the random variable that gives the distance from a specified end of the line will be some number in the interval [0, 1].

To study a *discrete* random variable, list the specific values that the random variable could take along with their probabilities, as we did for the random variables involving dice in Chapter 7.

We can't do this for a continuous random variable because it can take on infinitely many different values. Instead, for a continuous random variable X, we can find the probability that X will take on a value in an interval, but not for a specific value. A function gives the distribution, and areas under the function correspond to probabilities, as we shall see. This function is called a **continuous probability function** or a **probability density function**.

The most important property of a probability density function is that the area under the graph of the function must be equal to 1. This corresponds to the property that the sum of all the probabilities for a discrete random variable is 1, which we saw earlier.

The graph of the density function for the random variable M that gives the distance from the left endpoint on a line 1 meter long of a point chosen at random is given in Figure 8.1. It is a straight horizontal line because any point along the line is equally likely. It has height equal to 1 since that will give an area equal to 1.

To find the probability that the random variable M will take on a value in a certain interval, we find the area under the line corresponding to the interval. So if we want to know the probability that the value of M is in the interval [0.2, 0.5], we look at the area shown in Figure 8.2. This area is $(0.5 - 0.2) \times 1 = 0.3$. This tells us the probability that M will be in the interval from 0.2 to 0.5 is 0.3, and we write $P\{0.2 \leq M \leq 0.5\} = 0.3$.

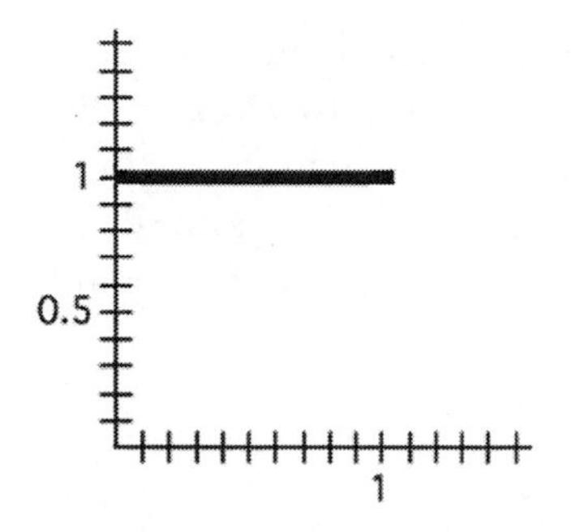

Figure 8.1
The probability density function of the random variable *M*.

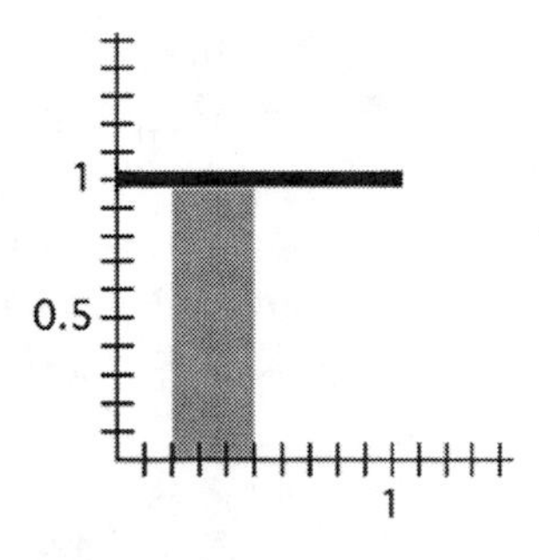

Figure 8.2
The shaded area corresponds to the probability that the random value *M* will take on a value in the interval [0.2, 0.5].

Sometimes, a discrete random variable can take on many different values or very large values. In such cases, even though there are a finite number of different outcomes for the random trials, it is easier to use a continuous random variable and a probability density function rather than a discrete random variable with a list of values. Common situations that are modeled with continuous random variables are large populations and large amounts of money. In other situations with many possible values or very large values, you have to use your own best judgment.

8.2 Cumulative Distribution Function of a Continuous Random Variable

As we have seen for discrete random variables, the cumulative distribution function of a random variable is a useful way to help us understand the properties of the random variable. For a given value x of the random variable, the cumulative distribution function gives the probability that the random variable takes on a value up to and including the given value x. In symbols, the cumulative distribution function for the random variable X is the function $F(x) = \mathrm{P}\{X \leq x\}$.

The cumulative distribution function (or CDF) for a continuous random variable is, like the probability density function, often shown as a graph or given as an equation. The graph of a cumulative distribution function shows how the probability increases for greater and greater values of the random variable. The graph of the CDF $F(x)$ for the random variable M described in section 8.1 is shown in Figure 8.3.

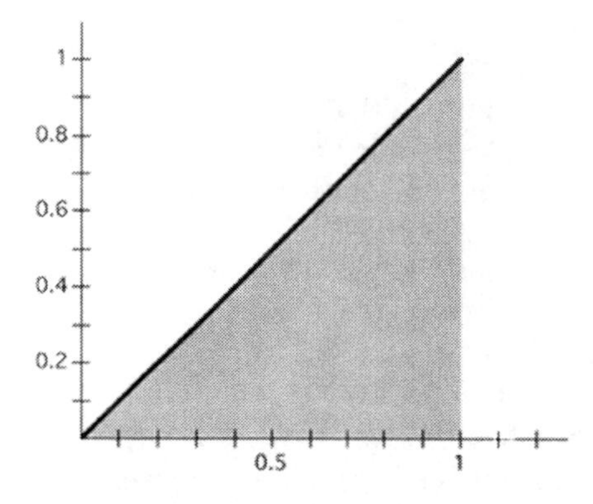

Figure 8.3
The cumulative distribution function of the random variable *M*.

Cumulative distribution functions for continuous random variables have important properties similar to those for discrete random variables. For a cumulative distribution function $F(x)$ belonging to the random variable X and numbers a and b:

- The function $F(x)$ keeps getting larger and is a nondecreasing function. In symbols, if $a < b$, then $F(a) \leq F(b)$.
- The value of $F(x)$, as a probability, is between 0 and 1. In symbols, $0 \leq F(x) \leq 1$.
- To find the probability that the random variable X takes on a value in the interval $(a, b]$, use the formula $P\{a < X \leq b\} = F(b) - F(a)$.
- To compute the value $F(a)$ if we know the probability density function, find the area under the probability density function from 0 to a.

If we want to compare two different probability distributions, look at the graphs of their CDFs plotted on the same grid. In Figure 8.4, there are two different CDFs that we can compare. The CDF represented by the dotted line increases more rapidly at first, showing that smaller values are more likely for the random variable that it represents. In fact, most values of this random variable will be between 0 and 0.5. The dashed line increases more slowly at first, showing that larger values are more likely for the corresponding random variable. Its largest values are a little greater than 2.

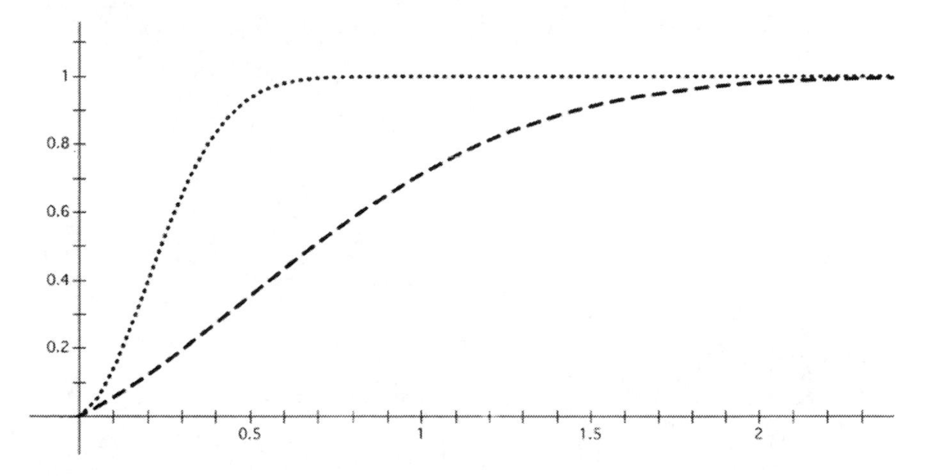

Figure 8.4
Two different cumulative distribution functions shown together.

8.3 The Uniform Distribution

Sometimes, all values of a random value are equally likely. In the example we saw before of choosing a point at random along a line one meter long and measuring its distance from an endpoint, each value was equally likely. A random variable like this, with outcomes falling along an interval and with every outcome equally likely, is said to have a **uniform distribution**.

The probability density function is a constant function and the cumulative distribution function is a straight line. This is the simplest continuous distribution.

Example of a Uniform Distribution

A uniform distribution is shown in Figure 8.5; call the random variable with this distribution W. The random variable takes on values in the interval from 0 to 2, and any value in this interval has the same likelihood. The probability density function has value equal to 0.5 on the interval [0, 2], so the area under the graph of the function is equal to 1. The graph of the CDF is a straight line from the origin reaching a height of 1 when $x = 2$. The probability $P\{1 < W \leq 2\} = 0.5$, since that is the area under the graph of the density function from $x = 1$ to $x = 2$. The probability $P\{W \leq 1.5\}$ is 0.75, since 0.75 is the value of the CDF for $x = 1.5$.

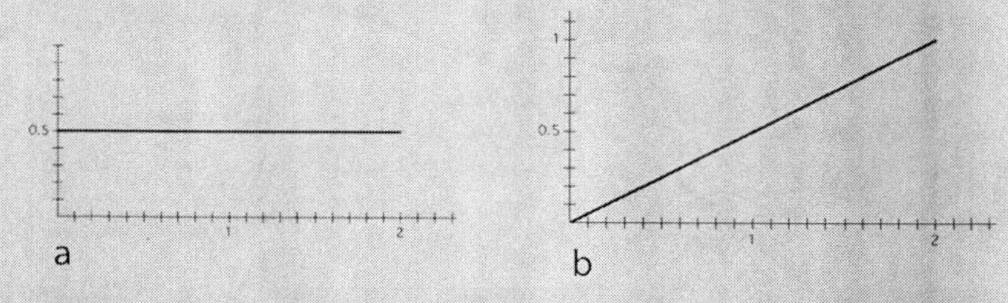

Figure 8.5
A uniform probability density function (a) and its cumulative density function (b).

8.4 The Normal Distribution

The normal random variable is the most important continuous random variable. Many measurements like height and weight are usually distributed normally. Most of the measurements are close to a mean (or average) value, with some much smaller and some much larger. The graph in Figure 8.6 shows a normal distribution with mean equal to 2.

This graph illustrates the most important properties of the normal distribution:

- The graph is symmetrical about its mean or average value.
- The graph tapers off gradually to values closer and closer to 0 on both sides.
- The overall shape of the graph is like a bell; for this reason, the graph is often called "the bell-shaped curve."

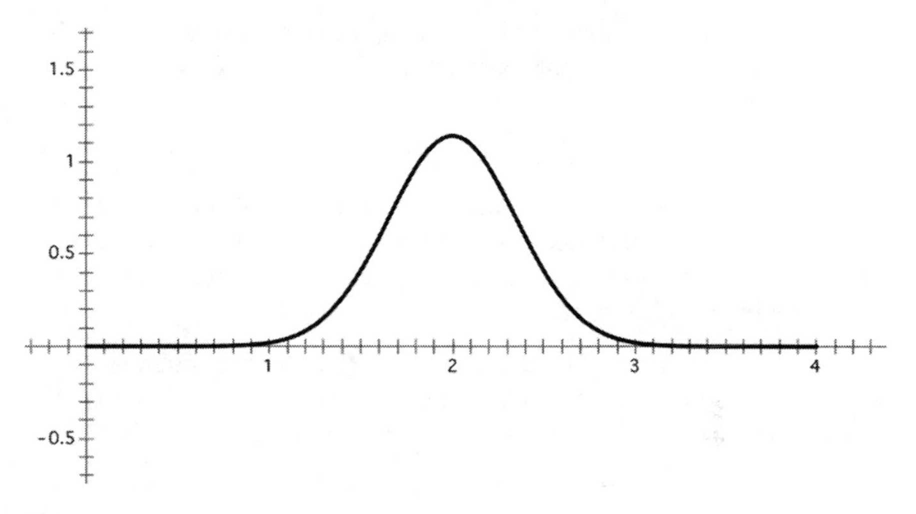

Figure 8.6
A normal probability density function with mean equal to 2.

Examples of random variables that can be modeled by a normal distribution:

- The weight of a robin's egg
- The height of an 18-year-old boy
- The error in a specific measurement
- The lifetime of a battery
- The score on a final exam

The normal distribution is also called the **Gaussian distribution**. Chapter 9 will give a more detailed discussion of this distribution and its applications.

8.5 The Exponential Distribution

The time until something happens can seem random. The time it takes someone ahead of you at the ATM to complete their transaction, the time you wait on hold for the help desk, the lifetime of a hard drive, and the time between two clicks on a Geiger counter all behave in a random way.

You cannot predict the time until one of these random events happen, but you can model them using the exponential distribution. This is given by the formula

$$f(t) = \lambda e^{-\lambda t} \text{ for } t \geq 0$$

where λ is a positive number, or parameter, that depends on the given situation. Instead of the usual variable x, the variable t is used for this distribution and represents time, since in most applications the random variable corresponds to a period of time.

The Poisson distribution, studied in Chapter 7, models the number of random events in a time interval, while the exponential distribution models the time until the next random event happens.

The graph in Figure 8.7 shows the exponential distribution function for several different values of λ.

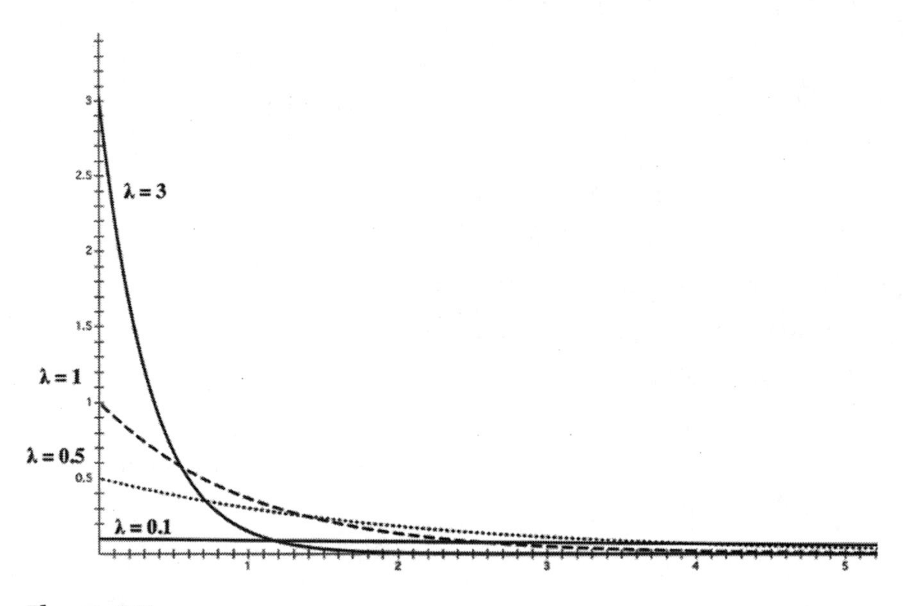

Figure 8.7
The exponential distribution for values of λ = 0.1, 0.5, 1, and 3.

There are several important features of the exponential distribution function that we can see from this figure.

- As t gets larger, the value of the function gets closer and closer to 0.
- The distribution function is always decreasing.
- The distribution decreases more rapidly at first.
- For smaller values of λ, the function starts off lower. This means that the probability of a lower value for the random variable is less likely when λ is smaller.
- For larger values of λ, the function starts off higher. This means that the probability of a lower value for the random variable is more likely when λ is larger.

The cumulative distribution function for the exponential distribution is given by the formula

$$F(t) = \mathrm{P}\{X \le t\} = 1 - e^{\lambda t}$$

The cumulative distribution function for several different values of λ is shown in Figure 8.8.

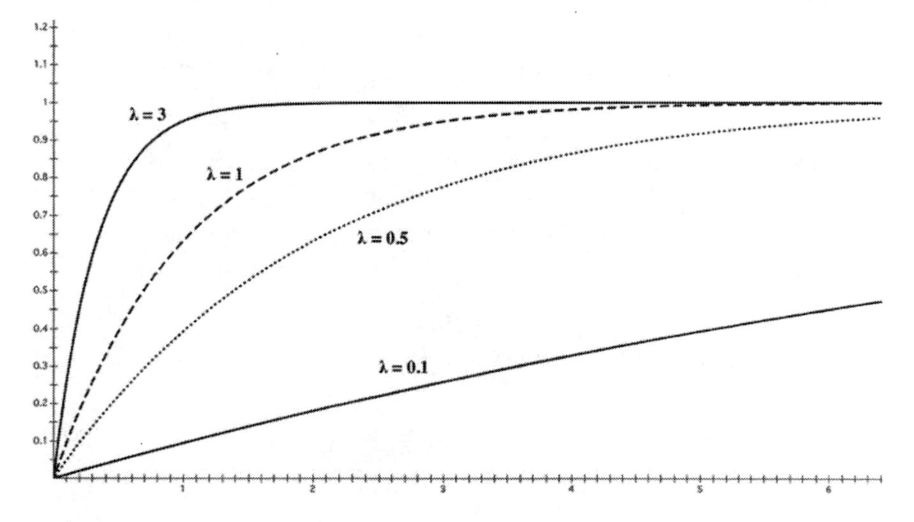

Figure 8.8
The exponential distribution for values of λ = 0.1, 0.5, 1, and 3.

The most important feature of the cumulative distribution function that we can see from this figure is that for larger values of λ, the CDF increases more rapidly at first and gets close to 1 more quickly than for smaller values of λ.

Examples of situations that can be modeled with an exponential distribution:

- The time until a light bulb burns out
- The time until the next customer appears at a bank
- The time until an amoeba divides into two

Examples of the Exponential Distribution

- Suppose that the time until a certain type of muffler wears out is modeled by the probability distribution function $f(t) = 0.2e^{-0.2t}$ and the cumulative distribution function $F(t) = \text{P}\{M \leq t\} = 1 - e^{0.2t}$, where t is given in years. The probability that a muffler will wear out in less than five years is $\text{P}\{M \leq 5\} = 1 - e^{-0.2 \times 5} \approx 0.632$. The probability that the muffler will last more than ten years is

$$\begin{aligned} \text{P}\{M > 10\} &= 1 - \text{P}\{M \leq 10\} = 1 - (1 - e^{-0.2 \times 10}) \\ &= e^{-0.2 \times 10} \approx 0.135 \end{aligned}$$

- Suppose that the lifetime of a new type of projector lamp is exponential with probability distribution function $f(t) = 0.03e^{-0.03t}$, where t is given in hours. Then its cumulative distribution function is

$$F(t) = \text{P}\{C \leq t\} = 1 - e^{-0.03t}$$

The probability that a projector lamp of this type will last at least 50 hours is

$$\begin{aligned} \text{P}\{C \geq 50\} &= 1 - \text{P}\{C \leq 50\} = 1 - (1 - e^{-0.03 \times 50}) \\ &= e^{-0.03 \times 50} \approx 0.223 \end{aligned}$$

If you expect to use the projector for more than 50 hours in the next month, you should have a replacement bulb on hand.

- The lifetime of a cobalt 56 atom has probability distribution function $f(t) = 0.009e^{-0.009t}$, where t is given in days. The cumulative distribution function is

$$F(t) = \mathrm{P}\{L \leq t\} = 1 - e^{-0.009t}$$

If you have one atom of cobalt 56, the probability that it will have decayed after 77 days is

$$\mathrm{P}\{L \leq 154\} = 1 - e^{-0.009 \times 77} \approx 1 - 0.50 \approx 0.05$$

We see that the probability of an atom of cobalt 56 decaying in a 77-day period is 50%, so we say that the **half-life** of cobalt 56 is 77 days.

The probability that one atom of cobalt 56 will have decayed after 154 days is

$$\mathrm{P}\{L \leq 154\} = 1 - e^{-0.009 \times 154} \approx 1 - 0.25 \approx 0.75$$

This means that after $154 = 2 \times 77$ days, the probability of decaying is about $0.50 \times 0.50 = 0.25$, so the probability of decay is $1 - 0.25 \approx 0.75$.

Chapter 8 Summary

A **continuous random variable** is a random variable with values anywhere in an interval.

A function called a **continuous probability function** or a **probability density function** gives the distribution of a continuous random variable. The most important property of a probability density function is that the area under the graph of the function must be equal to 1.

The cumulative distribution function for the random variable X is the function $F(x) = \mathrm{P}\{X \leq x\}$.

For numbers a and b, the important properties of a cumulative distribution function $F(x)$ belonging to the random variable X are as follows:

- The function $F(x)$is a **nondecreasing** function. So if $a < b$, then $F(a) \leq F(b)$.
- $0 \leq F(x) \leq 1$.
- To find the probability that the random variable X takes on a value in the interval $(a, b]$, we use the formula $P\{a < X \leq b\} = F(b) - F(a)$.
- To compute the value $F(a)$ if we know the probability density function, find the area under the probability density function from 0 to a.

A random variable with the property that every outcome is equally likely has a **uniform distribution**.

The **normal random variable** is used for measurements like height and weight. The most important properties of the normal distribution are as follows:

- The graph is symmetrical about its mean or average value.
- The graph tapers off gradually to values closer and closer to 0 on both sides.
- The overall shape of the graph is like a bell.

The **exponential distribution** is used to estimate the time until a random event happens and is given by the formula $f(t) = \lambda e^{-\lambda t}$ for $t \geq 0$, where λ is a positive number, or parameter, that depends on the given situation.

The important properties of the exponential distribution are as follows:

- As t gets larger, the value of the distribution function gets closer and closer to 0.
- The distribution function is always decreasing.

The **cumulative distribution function for the exponential distribution** is given by the formula $F(t) = P\{X \leq t\} = 1 - e^{\lambda t}$.

Chapter 8 Practice Problems

1. For each of the following random variables, tell whether it is discrete or continuous.
   a. The age of a student chosen at random
   b. The length of the right foot of a student chosen at random
   c. The shoe size of a student chosen at random
   d. The number of pets owned by a student chosen at random
   e. The weight of a hamster chosen at random
   f. The number of eggs in a bird's nest chosen at random
   g. The weight of a bird's egg chosen at random
   h. The number of bees in a hive chosen at random
   i. The honey production of a hive chosen at random
   j. The number on a card drawn at random from a standard deck of 52 (count a face card as 10)
   k. The population of a city chosen at random from all the cities in a state
   l. The income of a family chosen at random in a city
   m. Your winnings from an afternoon of rummy at $1.00 per game
   n. The lifetime of a light bulb chosen at random
   o. The number of letters in an e-mail
   p. The number of letters in a Tweet
2. For the graphs of the functions shown in Figures 8.9, 8.10, and 8.11, tell whether or not it could be the graph of a probability density function. Explain.

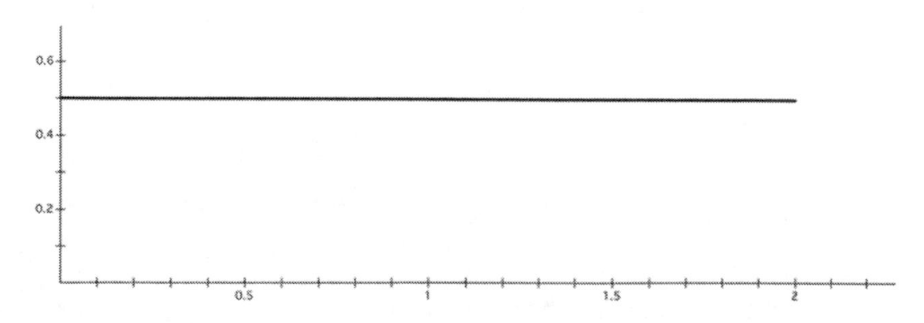

Figure 8.9
a. Could this be a probability density function?

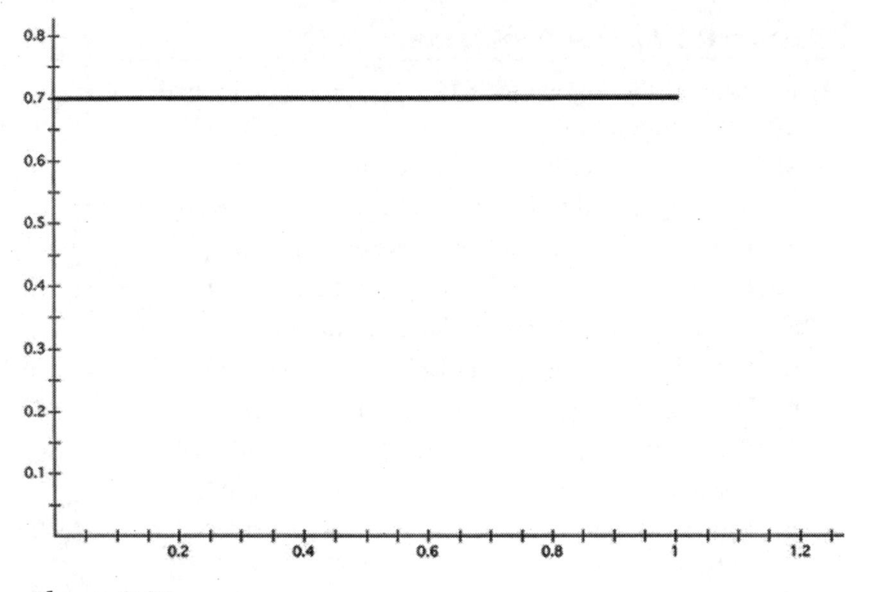

Figure 8.10
b. Could this be a probability density function?

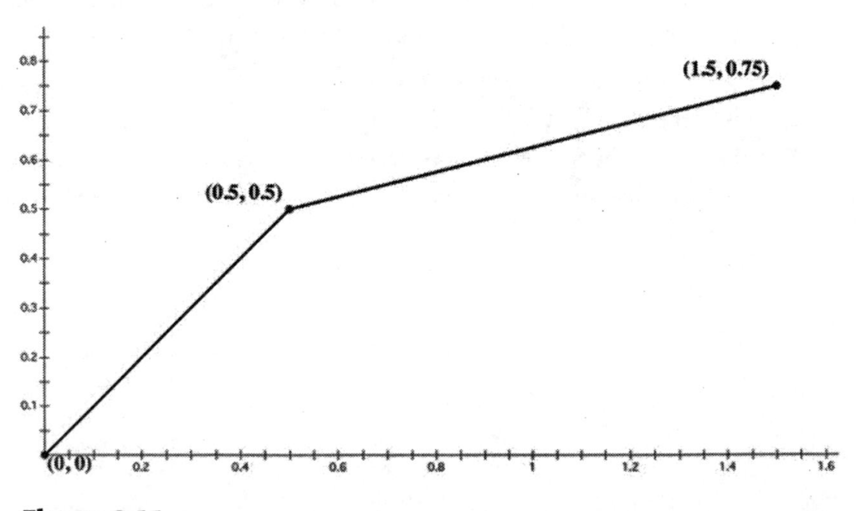

Figure 8.11
c. Could this be a probability density function?

3. Which of the graphs shown in Figures 8.12, 8.13, and 8.14 could belong to the cumulative distribution function of a continuous random variable? Explain.

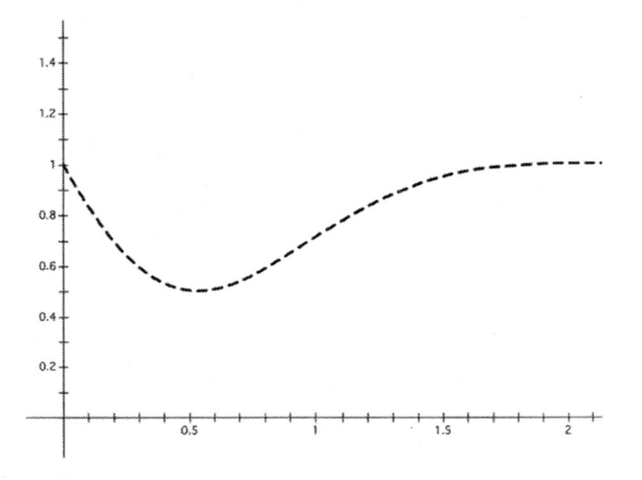

Figure 8.12
a. Could this be a cumulative distribution function?

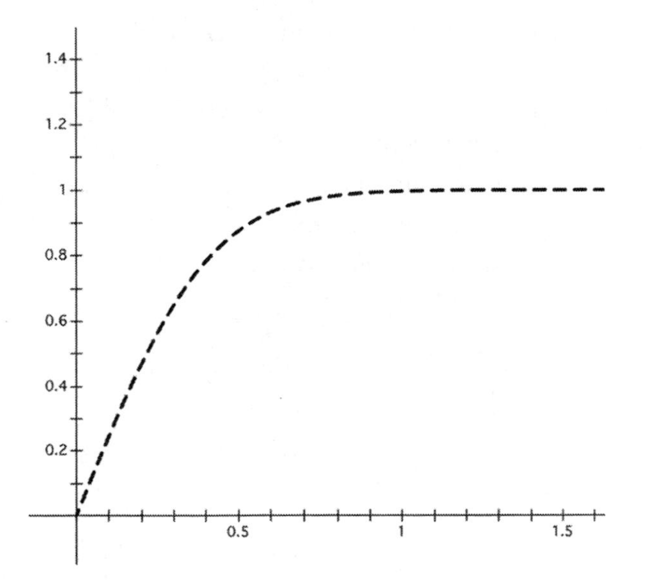

Figure 8.13
b. Could this be a cumulative distribution function?

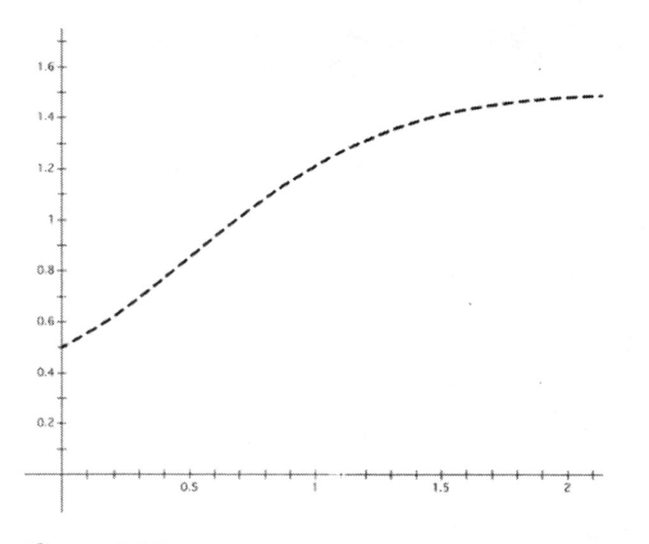

Figure 8.14

c. Could this be a cumulative distribution function?

4. Match each probability density function on the left in Figure 8.15 with the correct cumulative density function on the right.
5. Answer the following questions for the cumulative distribution function whose graph is shown in Figure 8.16.
   - **a.** Describe the properties of the continuous random variable Z whose cumulative distribution function is shown in Figure 8.16.
   - **b.** For this cumulative distribution function, what is $P\{Z \leq 0.4\}$?
   - **c.** For this cumulative distribution function, what is $P\{Z \leq 1.2\}$?
   - **d.** For this cumulative distribution function, what is $P\{Z \leq 1.6\}$?
   - **e.** What is $P\{1.6 \leq Z \leq 2.6\}$?
   - **f.** What is $P\{0.5 \leq Z \leq 1\}$?
   - **g.** What is $P\{Z > 1)$?

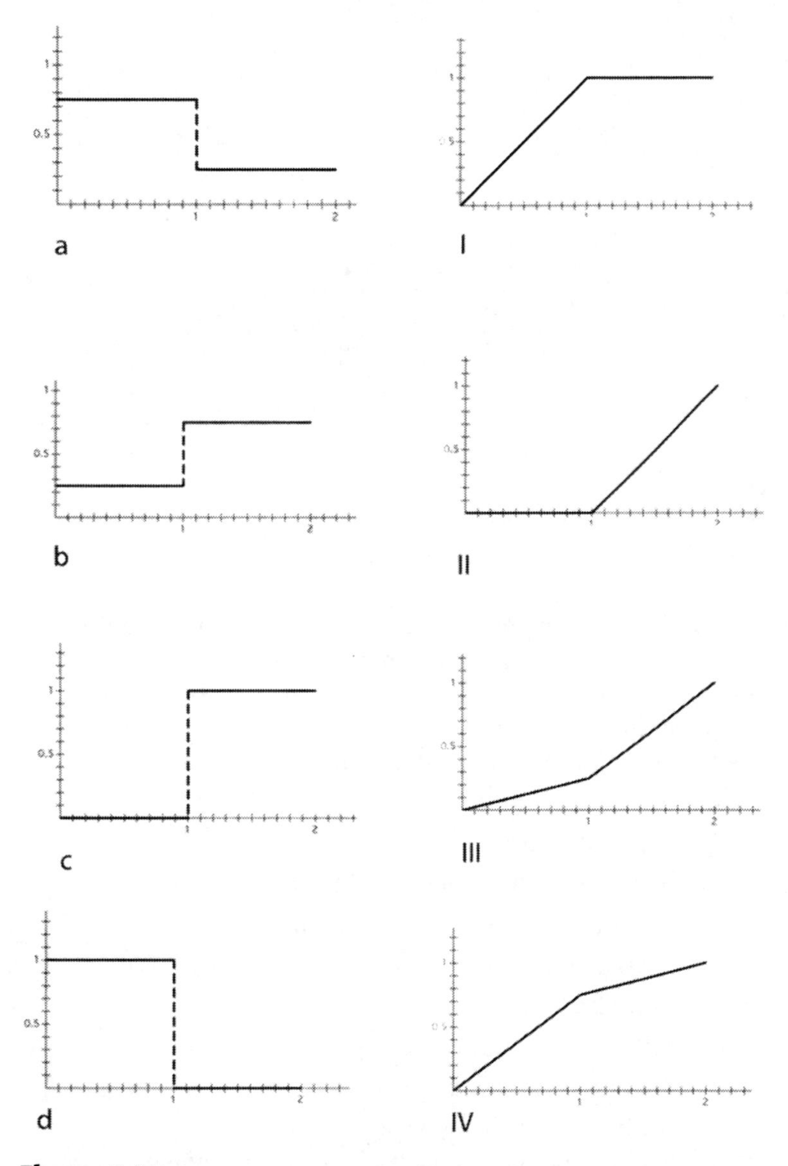

Figure 8.15
Match each probability density function on the left with the correct cumulative density function on the right.

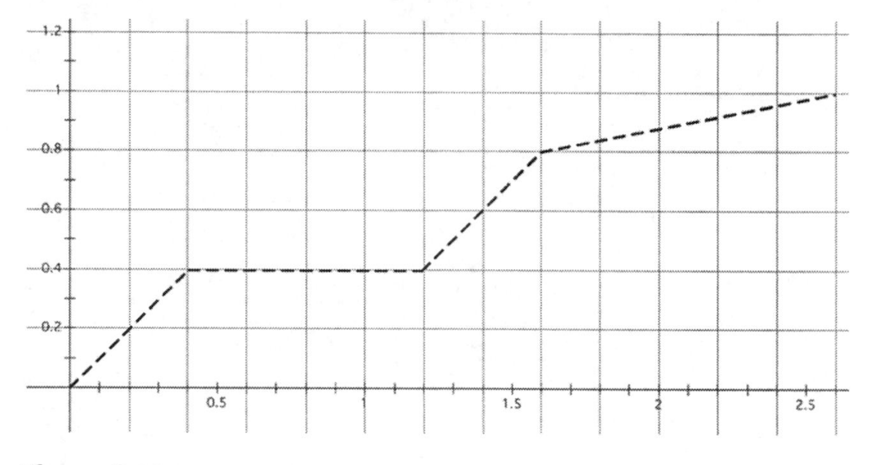

Figure 8.16
A cumulative distribution function.

6. For each of the following situations, choose which probability distribution is appropriate to use from this list: binomial distribution, Poisson distribution, geometric distribution, negative binomial distribution, hypergeometric distribution, uniform distribution, normal distribution, exponential distribution, or none of these.
   - **a.** The weight of a battery
   - **b.** The lifetime of a battery
   - **c.** How much longer this battery will last
   - **d.** The time until you see the next shooting star
   - **e.** The temperature of a day during the month of April for a given city
   - **f.** The length of a catfish in a pond
   - **g.** The length of a call made to a help desk
   - **h.** How much time you need to wait to be served at a help desk
   - **i.** Defective light bulbs in a package of 10
   - **j.** Defective light bulbs in a production run of 1,000
   - **k.** The number of coffee shops in a mall
   - **l.** The daily revenue of a coffee shop
   - **m.** The number you get at the deli
   - **n.** The probability of rain tomorrow
   - **o.** Throwing dice until you get doubles

p. Throwing dice until you get doubles three times
q. Throwing 10 dice and counting how many of the number 2 come up
r. Throwing 10 dice and getting three 3s and four 4s and three 7s
s. The number of rabbit holes in a field
t. The weights of the rabbits measured by a biologist
u. Capturing rabbits until you get a tagged rabbit
v. Capturing rabbits until you get 10 tagged rabbits
w. The number of tagged rabbits in a sample of 10 rabbits
x. The number of people in a city over the age of 100
y. The ages of people in a city
z. The incomes of families in a city

7. Figure 8.17 shows the graphs of six different probability density functions. For each, tell what kind of distribution it is: uniform, normal, or exponential.

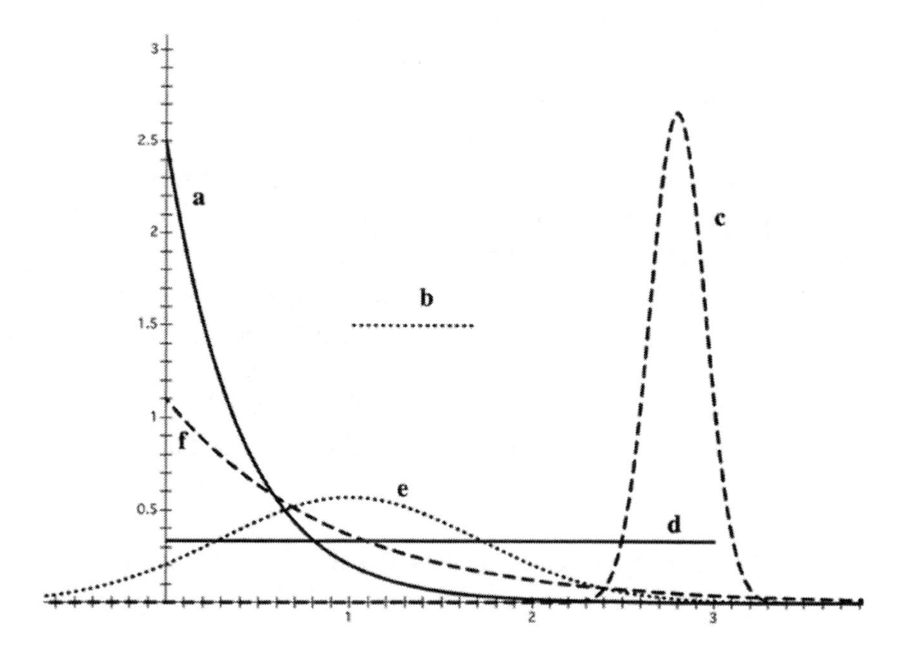

Figure 8.17
Six distribution functions.

Answers to Chapter 8 Practice Problems

1. **a.** Discrete (the age is given in years)
   b. Continuous (measurement)
   c. Discrete (there are only a few different shoe sizes)
   d. Discrete (the number will be an integer that is not very large, unless someone has an ant farm)
   e. Continuous (measurement)
   f. Discrete (the number will be relatively small)
   g. Continuous (measurement)
   h. Continuous (very large number)
   i. Continuous (measurement)
   j. Discrete
   k. Continuous (very large number)
   l. Continuous (money)
   m. Discrete (Even though this is money, the number of dollars won is an integer.)
   n. Continuous (measurement)
   o. Continuous (very large number of letters)
   p. Discrete if you think 140 is a small number of letters, or continuous if you think 140 is a large number.
2. **a.** Yes, because the area under the curve is $0.5 \times 2 = 1$.
   b. No, because the area under the curve is $0.7 \times 1 = 0.7$.
   c. No, because the area under the curve is $0.5 \times (0.5 \times 0.5) + 1 \times (0.5 + .75) / 2 = 0.125 + 0.625 = 0.75$. (Use the formulas for the area of a triangle and the area of a trapezoid.)
3. **a.** No, because the function shown in the graph is sometimes decreasing.
   b. Yes.
   c. No, because the function takes on values greater than 1.

4. **a.** IV
   b. III
   c. II
   d. I
5. **a.** The continuous random variable takes on values between 0 and 2.6. Since the graph is horizontal between 0.4 and 1.2, the random variable doesn't take on any values in this interval.
   b. 0.4
   c. 0.4
   d. 0.8
   e. $1 - 0.8 = 0.2$
   f. $0.4 - 0.4 = 0$
   g. 0.6. (This event $Z > 1$ is the complement of the event $Z \leq 1$, and $P\{Z \leq 1\} = 0.4$, so $P\{Z > 1\} = 1 - P\{Z \leq 1\} = 1 - 0.4 = 0.6$.)
6. **a.** Normal
   b. Normal
   c. Exponential
   d. Exponential
   e. Normal
   f. Normal
   g. Normal
   h. Exponential
   i. Binomial distribution
   j. Poisson distribution
   k. Poisson distribution
   l. Normal distribution
   m. Uniform distribution
   n. None of these
   o. Geometric distribution
   p. Negative binomial distribution
   q. Binomial distribution

- **r.** Multinomial distribution
- **s.** Poisson distribution
- **t.** Normal distribution
- **u.** Geometric distribution
- **v.** Negative binomial distribution
- **w.** Binomial distribution
- **x.** Poisson distribution
- **y.** Normal distribution
- **z.** Normal distribution

7. **a.** Exponential
   - **b.** Uniform
   - **c.** Normal
   - **d.** Uniform
   - **e.** Normal
   - **f.** Exponential

Chapter

The Normal Distribution

The normal distribution, introduced in the previous chapter, is the most important continuous probability distribution because of its many applications:

- Modeling many different kinds of real-life situations
- Approximating the binomial distribution
- Theoretical applications, the most important of which is the central limit theorem

9.1 Properties of the Normal Distribution

The probability density function of the normal distribution is given by the equation

$$f(x) = \frac{1}{\sqrt{2\pi}\sigma} e^{-(x-\mu)^2/2\sigma^2}$$

where μ and σ are parameters. The graph in Figure 9.1 shows a probability density function with $\mu = 0$ and $\sigma = 1$; this is the **standard normal distribution.**

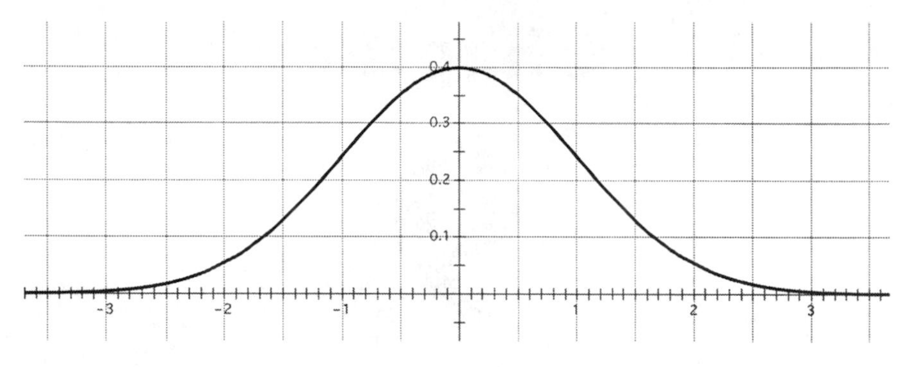

Figure 9.1
The density function of the standard normal distribution.

This graph illustrates several important features of the normal distribution, which we had observed in Chapter 8:

- The graph is symmetrical about its center.
- The graph tapers off gradually to values closer and closer to 0 on both sides.
- The area between the graph and the x-axis is equal to 1. The factor $\sqrt{2\pi}$ in the formula is needed to ensure that the area is 1.
- The overall shape of the graph is like a bell.

The parameters μ and σ give important information about the distribution and its graph.

The parameter μ is the **mean** of the distribution. This is a sort of average value—half of the area under the graph is to the right of the mean, and half of the area under the graph is to the left of the mean. The probability of getting a value below the mean is 0.5, and the probability of getting a value above the mean is also 0.5.

For a normal distribution function, the mean is always the largest value that the function takes on. For the standard normal distribution, the mean is 0, and the value of the distribution function at the mean is approximately 0.39894.

The graph in Figure 9.2 shows three normal distributions, each with a different mean and the same standard deviation. You can see that changing the mean moves the graph to the left or to the right. The highest point of the graph is directly above the mean. To estimate the mean from the graph of a normal distribution, look for the x-value of the highest point on the graph.

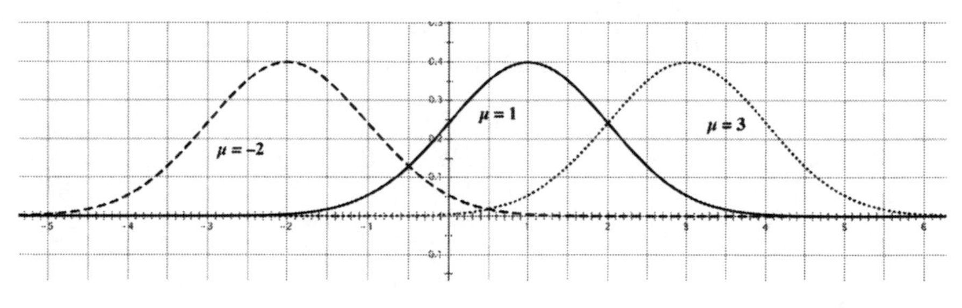

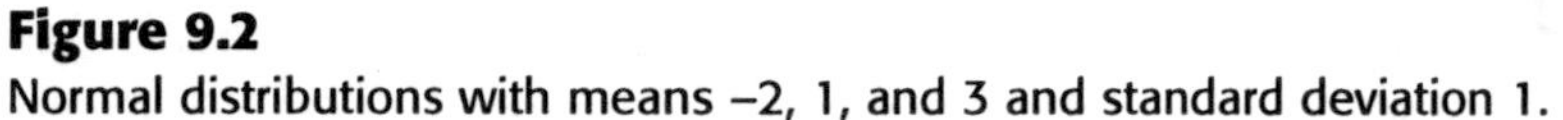

Figure 9.2
Normal distributions with means −2, 1, and 3 and standard deviation 1.

The parameter σ is the **standard deviation**, which tells how spread out the distribution is. The square of the standard deviation is called the **variance** and is denoted σ^2. The variance is a different measure of how spread out the distribution is.

The graph in Figure 9.3 shows three normal distributions, all with mean equal to 0, but each with a different standard deviation. As the standard deviation gets larger, the graph has a smaller value at the mean and appears to be spread out more.

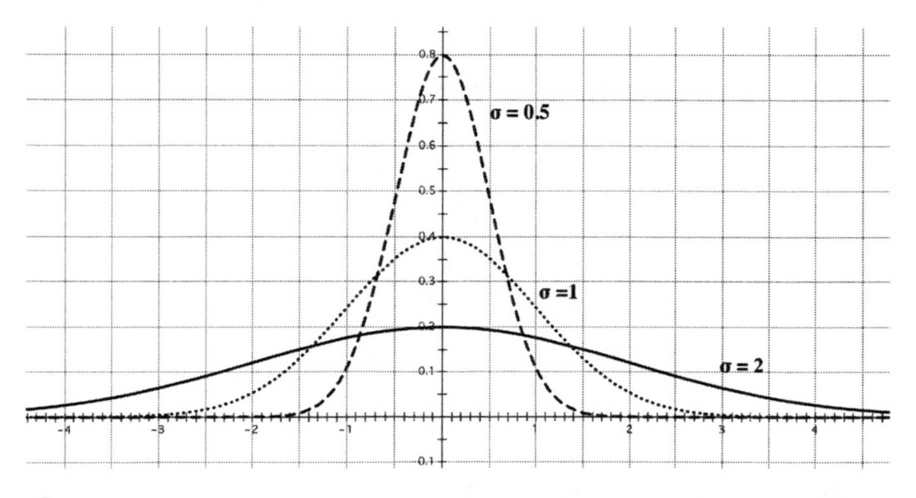

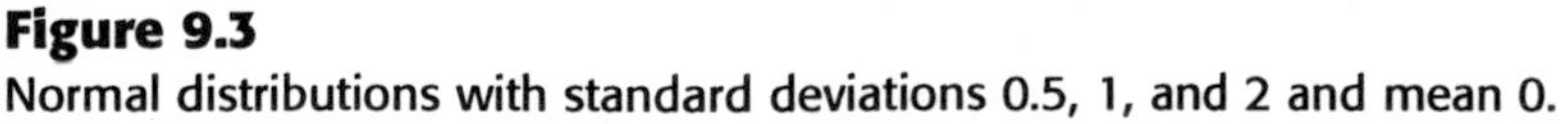

Figure 9.3
Normal distributions with standard deviations 0.5, 1, and 2 and mean 0.

A special feature of the graph of a normal distribution is that it has two inflection points. An **inflection point** is a point where concavity changes, either from concave down to concave up or from concave up to concave down. The two inflection points in the graph of the normal distribution make it look bell-shaped. Figure 9.4 shows a graph that is concave up, a graph that is concave down, and two graphs with inflection points.

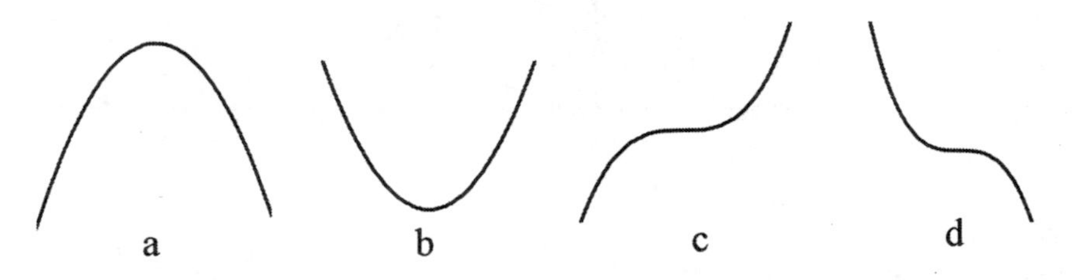

Figure 9.4
From left to right, a graph that is concave down (a), a graph that is concave up (b), and graphs with inflection points (c and d).

The graph of a normal distribution is concave up on the left, then concave down in the middle, and then concave up again on the right. There are two inflection points, one on either side of the mean μ and located at a distance equal to the standard deviation σ away from the mean, as shown in Figure 9.5.

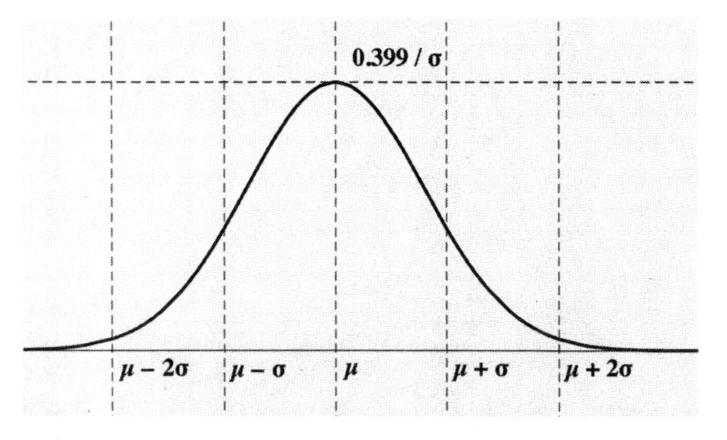

Figure 9.5

Normal distribution with standard deviation σ and mean μ.

The graph in Figure 9.5 shows that the standard deviation gives a measure of how spread out the graph of the normal distribution is. In fact, this spreading out is quite precise:

- Approximately 68.3% of the total area under the curve is between $\mu - \sigma$ and $\mu + \sigma$.
- Approximately 95.4% of the total area is between $\mu - 2\sigma$ and $\mu + 2\sigma$.
- Approximately 99.7% of the total area is between $\mu - 3\sigma$ and $\mu + 3\sigma$.

There are two ways to estimate the standard deviation σ from the graph of a normal distribution. You can locate one of the inflection points and see how far away it is from the mean; that distance is the standard deviation. Or you can estimate the maximum value that the function takes on (this happens when x is equal to the mean), set that number equal to $0.399 / \sigma$, and solve for σ.

9.2 Computing Probabilities for the Standard Normal Distribution

In this section, we will see how to compute probabilities for a random variable with normal distribution. In Chapter 8, we saw that the probability that a continuous random variable has a value in a given interval is equal to the area under the graph of the probability distribution function for that interval. Without using calculus, it is difficult to find the area under the standard normal distribution, but we can use a table to find probabilities for a standard normal distribution. For a normal distribution that is not standard, you will learn how to transform it in a way that lets you use the table for the standard normal distribution.

The random variable corresponding to the standard normal distribution is usually denoted by Z, while the random variable corresponding to any other normal distribution is denoted by X or some letter appropriate to the context.

The probability that the random variable Z takes on a value between a and b is written $P\{a \leq Z \leq b\}$ and corresponds to the area underneath the graph of the distribution function over the interval $[a, b]$. This area is shaded in Figure 9.6.

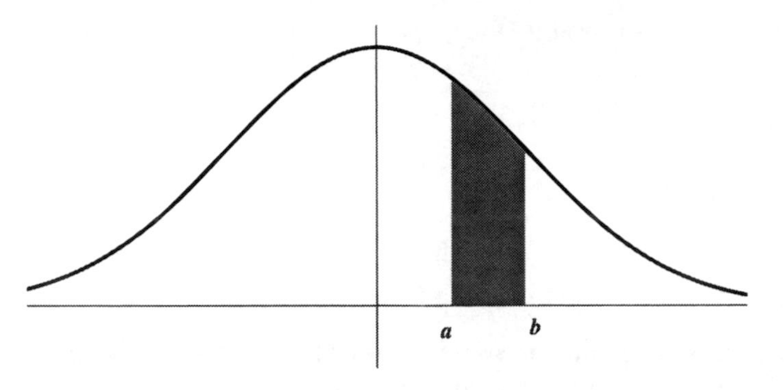

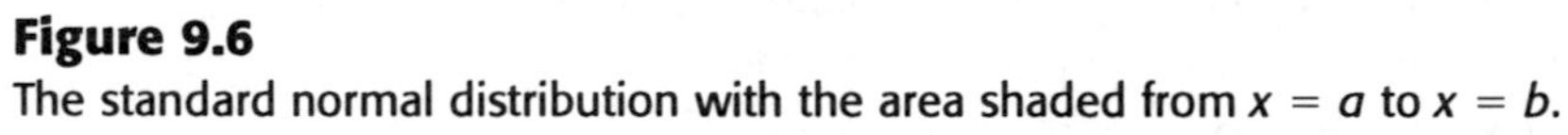

Figure 9.6
The standard normal distribution with the area shaded from $x = a$ to $x = b$.

Since it is very difficult to determine areas under a part of the standard normal distribution function, statisticians have computed tables that give numerical approximations for such areas. It is important to learn how to use these tables, which are very streamlined.

The area under the standard normal distribution graph from $x = 0$ to $x = z$ is called $\Phi(z)$. Thus $\Phi(z) = P\{0 \le X \le z\}$. Tables will typically give values for $\Phi(z)$ for non-negative z, that is, for $z \ge 0$. From these values and the properties of the standard normal distribution function, it is possible to compute areas for any other interval.

A table is given in Figure 9.7 for values of $\Phi(z) = P\{0 \le Z \le z\}$ with $0 \le z \le 3.99$, using steps of 0.01 for the values of z. For any value of z greater than 3.99, $\Phi(z)$ is approximately 0.5000.

It is easy to compute probabilities of the form $P\{0 \le Z \le z\}$ directly from the table, since $P\{0 \le Z \le z\} = \Phi(z)$. To compute other probabilities, you must use the properties of the graph of the normal distribution function. It is symmetric with respect to the y-axis: half of the area is to the left of the y-axis, half of the area is to the right of the y-axis, and the total area under the graph is equal to 1. It is always helpful to draw a sketch of the area that you need to find.

The area shaded in Figure 9.8 gives the probability that the random variable Z takes on a value in the interval $[0, a]$ on the x-axis, $P\{0 \le Z \le a\}$. This can be determined directly from the table and is simply $\Phi(a)$. For example, if we want to find the value of $P\{0 \le Z \le 0.32\}$, we get $\Phi(0.32) \approx 0.1255$ directly from the table in Figure 9.7.

z	$\Phi(z)$	z	$\Phi(z)$	z	$\Phi(z)$	z	$\Phi(z)$	z	$\Phi(z)$	z	$\Phi(z)$	z	$\Phi(z)$	z	$\Phi(z)$
0.01	0.0040	0.51	0.1950	1.01	0.3438	1.51	0.4345	2.01	0.4778	2.51	0.4940	3.01	0.4987	3.51	0.4998
0.02	0.0080	0.52	0.1985	1.02	0.3461	1.52	0.4357	2.02	0.4783	2.52	0.4941	3.02	0.4987	3.52	0.4998
0.03	0.0120	0.53	0.2019	1.03	0.3485	1.53	0.4370	2.03	0.4788	2.53	0.4943	3.03	0.4988	3.53	0.4998
0.04	0.0160	0.54	0.2054	1.04	0.3508	1.54	0.4382	2.04	0.4793	2.54	0.4945	3.04	0.4988	3.54	0.4998
0.05	0.0199	0.55	0.2088	1.05	0.3531	1.55	0.4394	2.05	0.4798	2.55	0.4946	3.05	0.4989	3.55	0.4998
0.06	0.0239	0.56	0.2123	1.06	0.3554	1.56	0.4406	2.06	0.4803	2.56	0.4948	3.06	0.4989	3.56	0.4998
0.07	0.0279	0.57	0.2157	1.07	0.3577	1.57	0.4418	2.07	0.4808	2.57	0.4949	3.07	0.4989	3.57	0.4998
0.08	0.0319	0.58	0.2190	1.08	0.3599	1.58	0.4429	2.08	0.4812	2.58	0.4951	3.08	0.4990	3.58	0.4998
0.09	0.0359	0.59	0.2224	1.09	0.3621	1.59	0.4441	2.09	0.4817	2.59	0.4952	3.09	0.4990	3.59	0.4998
0.10	0.0398	0.60	0.2257	1.10	0.3643	1.60	0.4452	2.10	0.4821	2.60	0.4953	3.10	0.4990	3.60	0.4998
0.11	0.0438	0.61	0.2291	1.11	0.3665	1.61	0.4463	2.11	0.4826	2.61	0.4955	3.11	0.4991	3.61	0.4998
0.12	0.0478	0.62	0.2324	1.12	0.3686	1.62	0.4474	2.12	0.4830	2.62	0.4956	3.12	0.4991	3.62	0.4999
0.13	0.0517	0.63	0.2357	1.13	0.3708	1.63	0.4484	2.13	0.4834	2.63	0.4957	3.13	0.4991	3.63	0.4999
0.14	0.0557	0.64	0.2389	1.14	0.3729	1.64	0.4495	2.14	0.4838	2.64	0.4959	3.14	0.4992	3.64	0.4999
0.15	0.0596	0.65	0.2422	1.15	0.3749	1.65	0.4505	2.15	0.4842	2.65	0.4960	3.15	0.4992	3.65	0.4999
0.16	0.0636	0.66	0.2454	1.16	0.3770	1.66	0.4515	2.16	0.4846	2.66	0.4961	3.16	0.4992	3.66	0.4999
0.17	0.0675	0.67	0.2486	1.17	0.3790	1.67	0.4525	2.17	0.4850	2.67	0.4962	3.17	0.4992	3.67	0.4999
0.18	0.0714	0.68	0.2517	1.18	0.3810	1.68	0.4535	2.18	0.4854	2.68	0.4963	3.18	0.4993	3.68	0.4999
0.19	0.0753	0.69	0.2549	1.19	0.3830	1.69	0.4545	2.19	0.4857	2.69	0.4964	3.19	0.4993	3.69	0.4999
0.20	0.0793	0.70	0.2580	1.20	0.3849	1.70	0.4554	2.20	0.4861	2.70	0.4965	3.20	0.4993	3.70	0.4999
0.21	0.0832	0.71	0.2611	1.21	0.3869	1.71	0.4564	2.21	0.4864	2.71	0.4966	3.21	0.4993	3.71	0.4999
0.22	0.0871	0.72	0.2642	1.22	0.3888	1.72	0.4573	2.22	0.4868	2.72	0.4967	3.22	0.4994	3.72	0.4999
0.23	0.0910	0.73	0.2673	1.23	0.3907	1.73	0.4582	2.23	0.4871	2.73	0.4968	3.23	0.4994	3.73	0.4999
0.24	0.0948	0.74	0.2704	1.24	0.3925	1.74	0.4591	2.24	0.4875	2.74	0.4969	3.24	0.4994	3.74	0.4999
0.25	0.0987	0.75	0.2734	1.25	0.3944	1.75	0.4599	2.25	0.4878	2.75	0.4970	3.25	0.4994	3.75	0.4999
0.26	0.1026	0.76	0.2764	1.26	0.3962	1.76	0.4608	2.26	0.4881	2.76	0.4971	3.26	0.4994	3.76	0.4999
0.27	0.1064	0.77	0.2794	1.27	0.3980	1.77	0.4616	2.27	0.4884	2.77	0.4972	3.27	0.4995	3.77	0.4999
0.28	0.1103	0.78	0.2823	1.28	0.3997	1.78	0.4625	2.28	0.4887	2.78	0.4973	3.28	0.4995	3.78	0.4999
0.29	0.1141	0.79	0.2852	1.29	0.4015	1.79	0.4633	2.29	0.4890	2.79	0.4974	3.29	0.4995	3.79	0.4999
0.30	0.1179	0.80	0.2881	1.30	0.4032	1.80	0.4641	2.30	0.4893	2.80	0.4974	3.30	0.4995	3.80	0.4999
0.31	0.1217	0.81	0.2910	1.31	0.4049	1.81	0.4649	2.31	0.4896	2.81	0.4975	3.31	0.4995	3.81	0.4999
0.32	0.1255	0.82	0.2939	1.32	0.4066	1.82	0.4656	2.32	0.4898	2.82	0.4976	3.32	0.4995	3.82	0.4999
0.33	0.1293	0.83	0.2967	1.33	0.4082	1.83	0.4664	2.33	0.4901	2.83	0.4977	3.33	0.4996	3.83	0.4999
0.34	0.1331	0.84	0.2995	1.34	0.4099	1.84	0.4671	2.34	0.4904	2.84	0.4977	3.34	0.4996	3.84	0.4999
0.35	0.1368	0.85	0.3023	1.35	0.4115	1.85	0.4678	2.35	0.4906	2.85	0.4978	3.35	0.4996	3.85	0.4999
0.36	0.1406	0.86	0.3051	1.36	0.4131	1.86	0.4686	2.36	0.4909	2.86	0.4979	3.36	0.4996	3.86	0.4999
0.37	0.1443	0.87	0.3078	1.37	0.4147	1.87	0.4693	2.37	0.4911	2.87	0.4979	3.37	0.4996	3.87	0.4999
0.38	0.1480	0.88	0.3106	1.38	0.4162	1.88	0.4699	2.38	0.4913	2.88	0.4980	3.38	0.4996	3.88	0.4999
0.39	0.1517	0.89	0.3133	1.39	0.4177	1.89	0.4706	2.39	0.4916	2.89	0.4981	3.39	0.4997	3.89	0.4999
0.40	0.1554	0.90	0.3159	1.40	0.4192	1.90	0.4713	2.40	0.4918	2.90	0.4981	3.40	0.4997	3.90	0.5000
0.41	0.1591	0.91	0.3186	1.41	0.4207	1.91	0.4719	2.41	0.4920	2.91	0.4982	3.41	0.4997	3.91	0.5000
0.42	0.1628	0.92	0.3212	1.42	0.4222	1.92	0.4726	2.42	0.4922	2.92	0.4982	3.42	0.4997	3.92	0.5000
0.43	0.1664	0.93	0.3238	1.43	0.4236	1.93	0.4732	2.43	0.4925	2.93	0.4983	3.43	0.4997	3.93	0.5000
0.44	0.1700	0.94	0.3264	1.44	0.4251	1.94	0.4738	2.44	0.4927	2.94	0.4984	3.44	0.4997	3.94	0.5000
0.45	0.1736	0.95	0.3289	1.45	0.4265	1.95	0.4744	2.45	0.4929	2.95	0.4984	3.45	0.4997	3.95	0.5000
0.46	0.1772	0.96	0.3315	1.46	0.4279	1.96	0.4750	2.46	0.4931	2.96	0.4985	3.46	0.4997	3.96	0.5000
0.47	0.1808	0.97	0.3340	1.47	0.4292	1.97	0.4756	2.47	0.4932	2.97	0.4985	3.47	0.4997	3.97	0.5000
0.48	0.1844	0.98	0.3365	1.48	0.4306	1.98	0.4761	2.48	0.4934	2.98	0.4986	3.48	0.4997	3.98	0.5000
0.49	0.1879	0.99	0.3389	1.49	0.4319	1.99	0.4767	2.49	0.4936	2.99	0.4986	3.49	0.4998	3.99	0.5000
0.50	0.1915	1.00	0.3413	1.50	0.4332	2.00	0.4772	2.50	0.4938	3.00	0.4987	3.50	0.4998	4.00	0.5000

Figure 9.7

Values of $\Phi(z) = P\{0 \le Z \le z\}$ for z between 0 and 3.99 with steps of 0.01.

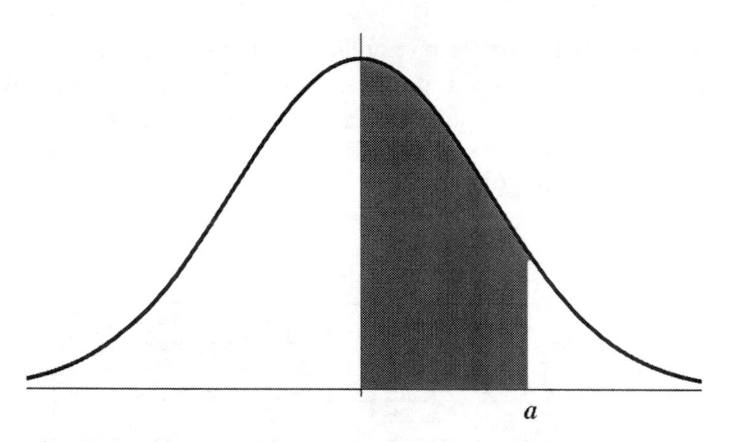

Figure 9.8

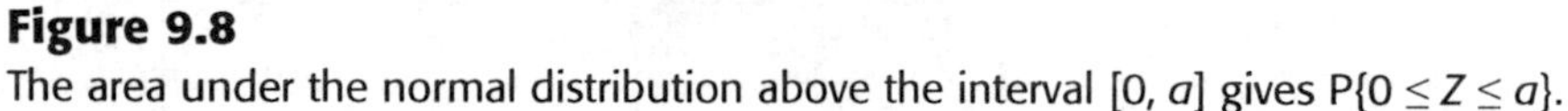

The area under the normal distribution above the interval $[0, a]$ gives $P\{0 \le Z \le a\}$.

Similarly, the area shaded in Figure 9.9 gives the probability that the random variable Z takes on a value in the interval $[-a, 0]$, $P\{-a \le Z \le 0\}$. Since the normal distribution curve is symmetric with respect to the y-axis, this area is the same as the area under the curve and above the interval $[0, a]$. So the probability $P\{-a \le Z \le 0\}$ is the same as the probability $P\{0 \le Z \le a\}$, which can be determined directly from the table. Thus $P\{-a \le Z \le 0\}$ is $\Phi(a)$. For example, the value of $P\{-0.32 \le Z \le 0\}$ is the same as $\Phi(0.32) \approx 0.1255$.

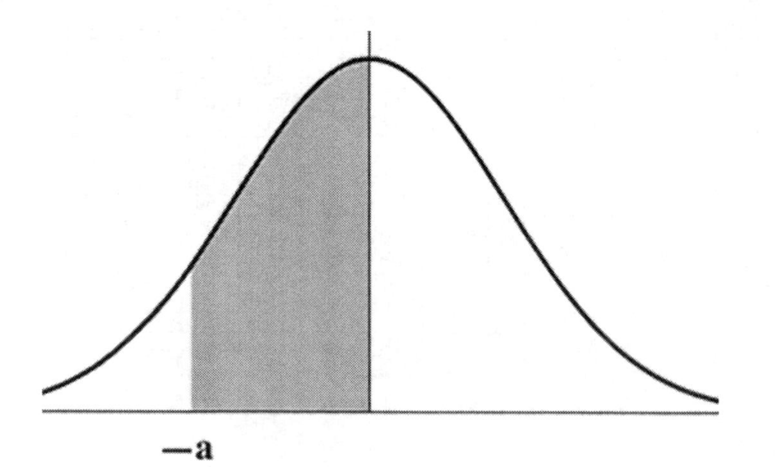

Figure 9.9

The area under the normal distribution above the interval $[-a, 0]$ gives $P\{-a \le Z \le 0\}$.

To find the probability that the random variable Z takes on a value in the interval $[-a, b]$, $P\{-a \leq Z \leq b\}$, you need to find the area under the normal distribution curve and above the interval $[-a, b]$ on the x-axis, shown in Figure 9.10. This area can be decomposed into two areas, the area above the interval $[-a, 0]$ and the area above the interval $[0, b]$. Thus $P\{-a \leq Z \leq b\} = P\{-a \leq Z \leq 0\} + P\{0 \leq Z \leq b\} = \Phi(a) + \Phi(b)$. So the value of $P\{-0.32 \leq Z \leq 1.27\}$ is the same as $\Phi(0.32) + \Phi(1.27) \approx 0.1255 + 0.3980 = 0.5235$.

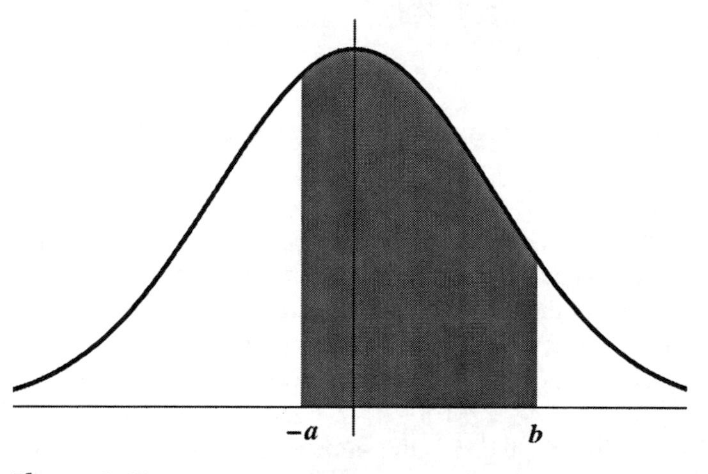

Figure 9.10
The area under the normal distribution above the interval $[-a, b]$ gives $P\{-a \leq Z \leq b\}$.

For the probability that the random variable Z takes on a value in the interval $[a, b]$, for any numbers a and b greater than 0, you need to find the area under the normal distribution curve that lies above the interval $[a, b]$ on the x-axis, as shown in Figure 9.11. This area can be understood as the difference of two areas, the area under the curve that lies above the interval $[0, b]$ minus the area under the curve that lies above the interval $[0, a]$. Thus $P\{a \leq Z \leq b\} = P\{0 \leq Z \leq b\} - P\{0 \leq Z \leq a\} = \Phi(b) - \Phi(a)$. So the value of $P\{0.52 \leq Z \leq 1.78\}$ is the same as $\Phi(1.78) - \Phi(0.52) \approx 0.4625 - 0.1985 \approx 0.2640$.

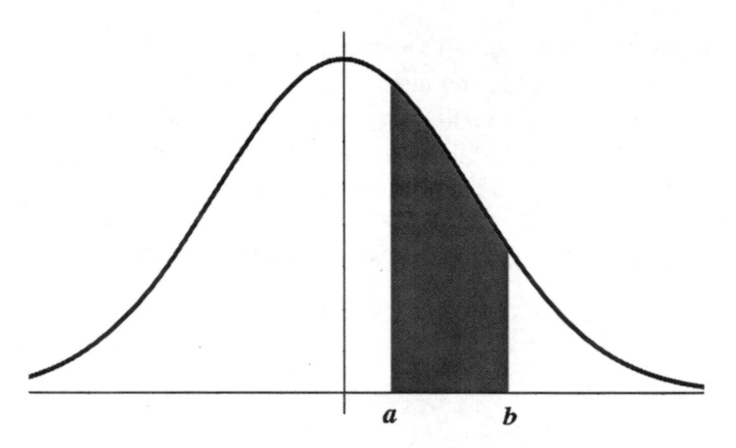

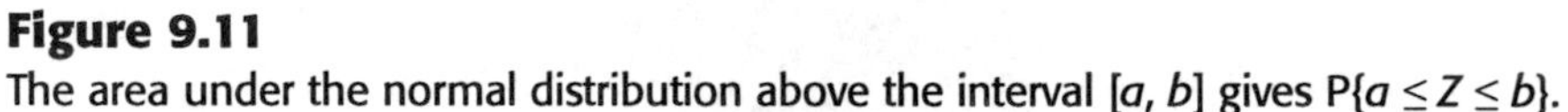

Figure 9.11
The area under the normal distribution above the interval $[a, b]$ gives $P\{a \leq Z \leq b\}$.

If now you want the probability that the random variable Z takes on a value in the interval $[-a, -b]$, which is $P\{-a \leq Z \leq -b\}$, you need to find the area under the normal distribution curve above the interval $[-a, -b]$ on the x-axis, as shown in Figure 9.12. Like the area above the interval $[a, b]$, this area can be understood as the difference of two areas, the area above the interval $[-a, 0]$ minus the area above the interval $[-b, 0]$. Thus $P\{-a \leq Z \leq -b\} = P\{-a \leq Z \leq 0\} - P\{-b \leq Z \leq 0\} = \Phi(a) - \Phi(b)$. So the value of $P\{-1.07 \leq Z \leq 0.55\}$ is the same as $\Phi(1.07) - \Phi(0.55) \approx 0.3577 + 0.2088 = 0.1489$.

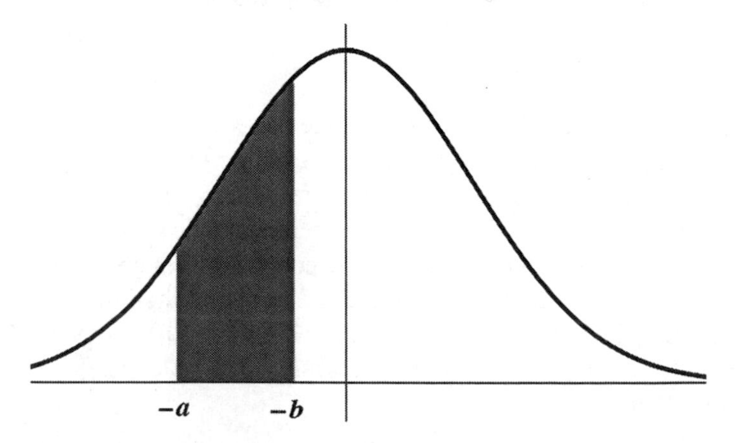

Figure 9.12
The area under the normal distribution above the interval $[-a, -b]$ gives $P\{-a \leq Z \leq -b\}$.

Sometimes, you want a probability that corresponds to an area over an infinite interval. For example, the probability $P\{a \leq Z\}$ that the random variable takes on a value greater than a corresponds to the area under the graph above the interval $[a, \infty]$. This is shown in Figure 9.13. Since we know that the area to the left of the y-axis is 0.5, we can subtract the area above the interval from $[0, a]$ from 0.5 to get the area above the interval $[a, \infty]$. Thus $P\{a \leq Z\} = 0.5 - \Phi(a)$. To compute $P\{1 \leq Z\}$, for example, we get $P\{1 \leq Z\} = 0.5 - \Phi(1) \approx 0.5 - 0.3413 \approx 0.1587$.

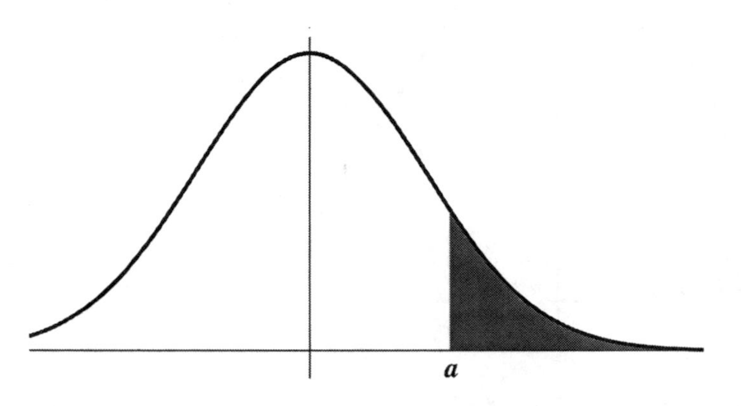

Figure 9.13
The area above the interval $[a, \infty]$ gives $P\{a \leq Z\}$ and is $0.5 - \Phi(a)$.

If the infinite interval is of the form $[-\infty, -a]$, you can use a similar method. For example, the probability $P\{Z \leq -a\}$ that the random variable takes on a value less than the negative number $-a$ corresponds to the area under the graph above the interval $[\infty, -a]$. This is shown in Figure 9.14. Since we know that the area to the left of the y-axis is 0.5, we can subtract the area above the interval $[-a, 0]$ from 0.5 to get the area above the interval $[\infty, -a]$. Remember that the area above the interval $[-a, 0]$ is the same as the area above the interval $[0, a]$. Thus $P\{Z \leq -a\} = 0.5 - \Phi(a)$. To compute $P\{Z \leq -1.25\}$, for example, we get $P\{Z \leq -1.25\} = 0.5 - \Phi(1.25) \approx 0.5 - 0.3944 \approx 0.1056$.

If you want the probability $P\{Z \leq a\}$ that the random variable takes on a value less than a positive number a, you need to find the area under the graph above the interval $[-\infty, a]$. This is shown in Figure 9.15. Since we know that the area to the left of the y-axis is 0.5, we can add the area above the interval from $[0, a]$ to 0.5 to get the area above the interval $[-\infty, a]$. Thus $P\{Z \leq a\} = 0.5 + \Phi(a)$. To compute $P\{Z \leq 0.51\}$, for example, we get $P\{Z \leq 0.51\} = 0.5 + \Phi(0.51) \approx 0.5 + 0.1950 = 0.6950$.

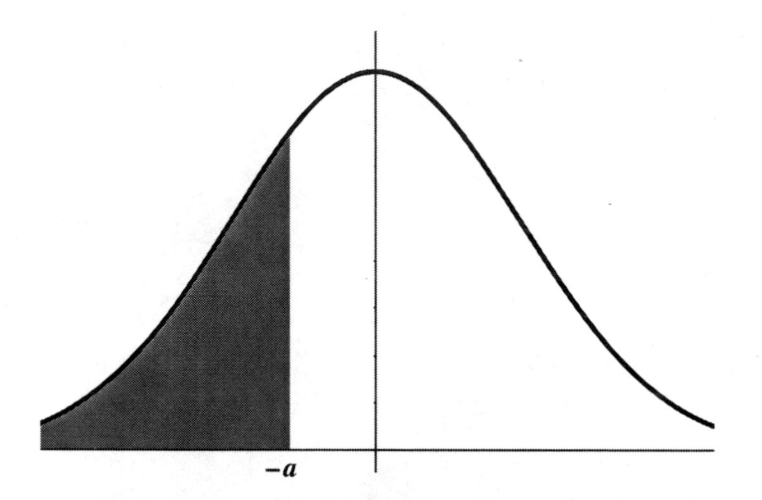

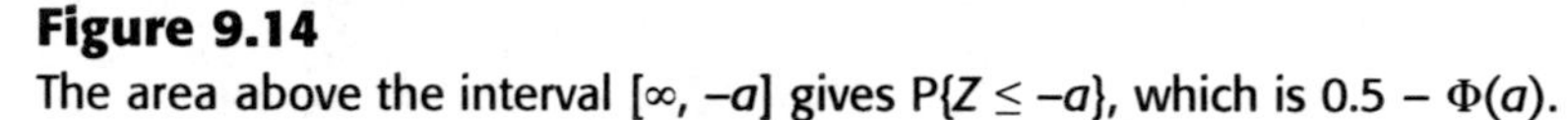

Figure 9.14
The area above the interval $[\infty, -a]$ gives $P\{Z \leq -a\}$, which is $0.5 - \Phi(a)$.

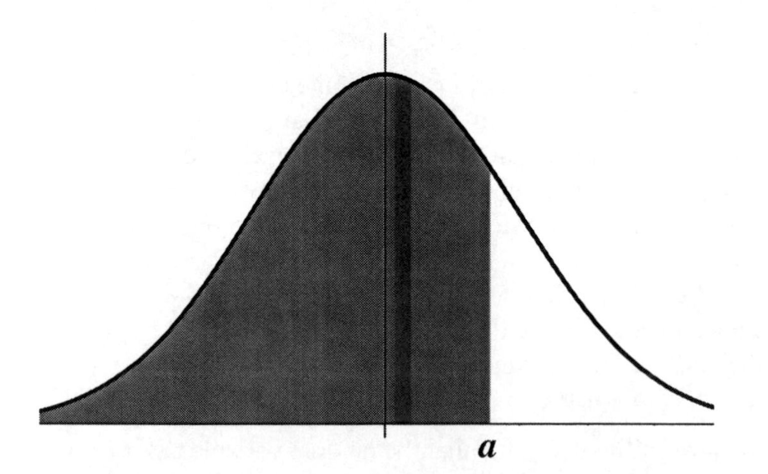

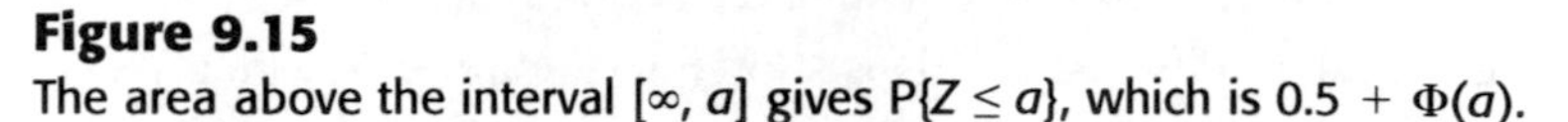

Figure 9.15
The area above the interval $[\infty, a]$ gives $P\{Z \leq a\}$, which is $0.5 + \Phi(a)$.

For probability $P\{-a \leq Z\}$ that the random variable takes on a value greater than a negative number $-a$, you need to compute the area under the graph above the interval $[-\infty, -a]$. This is shown in Figure 9.16. Since we know that the area to the right of the y-axis is 0.5 and the area above the interval $[-a, 0]$ is the same as $\Phi(a)$, we see that $P\{-a \leq Z\} = 0.5 + \Phi(a)$. To compute $P\{-0.68 \leq Z\}$, for example, we get $P\{-0.68 \leq Z\} = 0.5 + \Phi(0.68) \approx 0.5 + 0.2517 \approx 0.7517$.

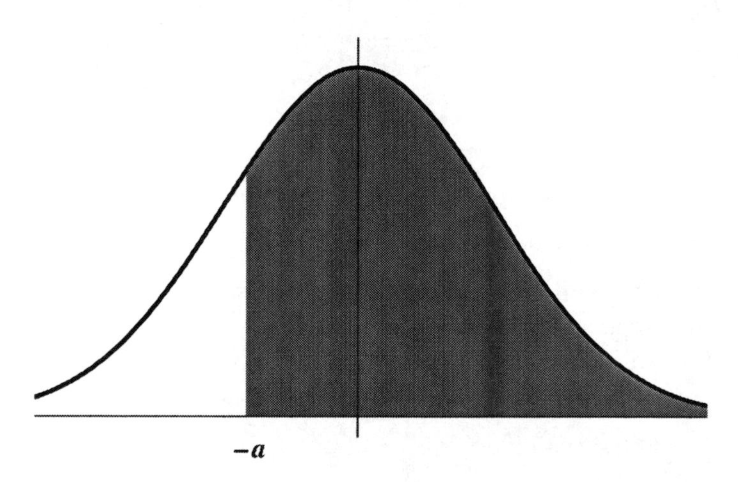

Figure 9.16
The area above the interval $[-a, \infty]$ gives $P\{Z \leq -a\}$, which is $0.5 - \Phi(a)$.

Examples of Computations for the Standard Normal Distribution

a. $P\{0 \leq Z \leq 2.15\} = \Phi(2.15) \approx 0.4842$. See Figure 9.8.

b. $P\{0 \leq Z \leq 1.03\} = \Phi(1.03) \approx 0.3485$. See Figure 9.8.

c. $P\{-1.4 \leq Z \leq 0\} = \Phi(1.4) \approx 0.4192$. See Figure 9.9.

d. $P\{-2.02 \leq Z \leq 1.5\} = \Phi(2.02) + \Phi(1.5) \approx 0.4783 + 0.4332 = 0.9115$. See Figure 9.10.

e. $P\{0.8 \leq Z \leq 0.9\} = \Phi(0.9) - \Phi(0.8) \approx 0.3159 - 0.2881 = 0.02780$. See Figure 9.11.

f. $P\{-2.5 \leq Z \leq -0.5\} = \Phi(2.5) - \Phi(0.5) \approx 0.4938 - 0.1915 = 0.3023$. See Figure 9.12.

g. $P\{1.01 \leq Z\} = 0.5 - \Phi(1.01) \approx 0.5 - 0.3438 = 0.1562$. See Figure 9.13.

h. $P\{Z \leq -1.4\} = 0.5 - \Phi(1.4) \approx 0.5 - 0.4192 = 0.08076$. See Figure 9.14.

i. $P\{Z \leq 0.2\} = 0.5 + \Phi(0.2) \approx 0.5 + 0.0793 = 0.5793$. See Figure 9.15.

j. $P\{-2 \leq Z\} = 0.5 + \Phi(2) \approx 0.5 + 0.4772 = 0.9772$. See Figure 9.16.

Summary for Computing Probabilities for the Standard Normal Distribution

When computing a probability for a standard normal distribution, it is helpful to draw the graph and mark the interval desired. From the graph, it is easy to see how to use the table to get the area needed. The summary given here can be also used as an aid in computing probabilities.

1. $P\{0 \leq Z \leq a\} = \Phi(a)$. See Figure 9.8.
2. $P\{-a \leq Z \leq 0\} = \Phi(a)$. See Figure 9.9.
3. $P\{-a \leq Z \leq b\} = \Phi(a) + \Phi(b)$. See Figure 9.10.
4. $P\{a \leq Z \leq b\} = \Phi(b) - \Phi(a)$. See Figure 9.11.
5. $P\{-a \leq Z \leq -b\} = \Phi(a) - \Phi(b)$. See Figure 9.12.
6. $P\{a \leq Z\} = 0.5 - \Phi(a)$. See Figure 9.13.
7. $P\{Z \leq -a\} = 0.5 - \Phi(a)$. See Figure 9.14.
8. $P\{Z \leq a\} = 0.5 + \Phi(a)$. See Figure 9.15.
9. $P\{-a \leq Z\} = 0.5 + \Phi(a)$. See Figure 9.16.

9.3 Computing Probabilities for Any Normal Distribution

If we have a normal distribution that is *not* standard, the table in Figure 9.7 can still be used to compute probabilities. In order to use the table, we must transform the normal distribution to the standard normal distribution, and then find the corresponding area, which will give the desired probability.

If X is a normal distribution with mean μ and standard deviation σ, we can transform it to the standard normal distribution Z by using the formula

$$z = \frac{x - \mu}{\sigma}$$

or

$$F(x) = \Phi\left(\frac{x - \mu}{\sigma}\right)$$

Then we compute a probability like $P\{a \leq X \leq b\}$ for the random variable X having normal distribution with mean μ and standard deviation σ by first finding

$$z_1 = \frac{a - \mu}{\sigma} \text{ and } z_2 = \frac{b - \mu}{\sigma}$$

and then computing $P\{z_1 \leq Z \leq z_2\}$ using the table for the standard normal distribution.

To find $P\{0.81 \leq X \leq 1.74\}$ for the normal distribution X with mean $\mu = -3$ and standard deviation $\sigma = 2$, first transform the numbers 0.81 and 1.74, and then use the table. Thus

$$z_1 = \frac{0.81 - (-3)}{2} = \frac{3.81}{2} = 1.905$$

and

$$z_2 = \frac{1.74 - (-3)}{2} = \frac{4.74}{2} = 2.37$$

Then $P\{0.81 \leq X \leq 1.74\} = P\{1.905 \leq Z \leq 2.37\} \approx \Phi(2.37) - \Phi(1.905) \approx 0.4911 - 0.4716 \approx 0.01950$.

The estimation for $\Phi(1.905)$ was done using the technique of linear interpolation, described in the next section.

Linear Interpolation

Linear interpolation is a technique for estimating a number from other numbers in a table. For example, in Figure 9.17, the numbers p and q correspond to numbers a and b in a table, and we want to find the number y between p and q that corresponds to the number x between the numbers a and b. Use the simple proportion

$$\frac{x-a}{b-a}=\frac{y-p}{q-p}$$

and solve for y to get

$$y=(q-p)\left(\frac{x-a}{b-a}\right)+p$$

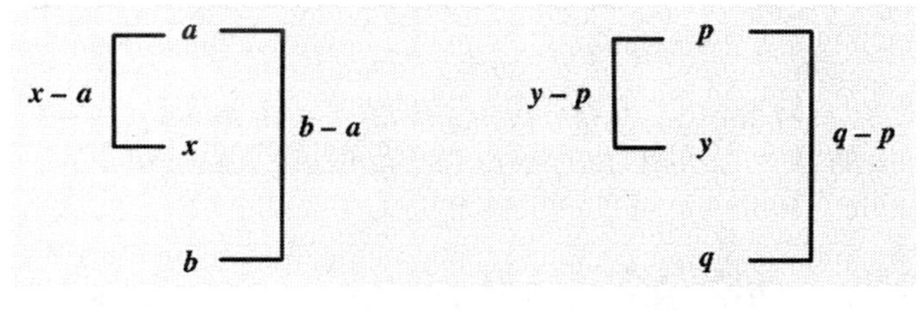

Figure 9.17
Numbers in two columns of a table along with differences.

For most tables of data, this method of interpolation only gives an approximate value.

We now use this method to compute $\Phi(1.905)$. From the table for the normal distribution shown in Figure 9.7, we see that $\Phi(1.90) \approx 0.4713$ and $\Phi(1.91) \approx 0.4719$. Then

$$\begin{aligned}\Phi(1.905) &\approx \big(\Phi(1.91)-\Phi(1.90)\big)\left(\frac{1.905-1.90}{1.91-1.90}\right)+\Phi(1.90)\\ &\approx \big(0.4719-0.4713\big)\left(\frac{0.005}{0.01}\right)+0.4713\\ &\approx (0.0006)(0.5)+0.4713\\ &\approx 0.4716\end{aligned}$$

Examples of Computing Normal Distributions

- Find P$\{0 \leq X \leq 1.28\}$ with $\mu = 2$ and $\sigma = 1.2$. Then

$$z_1 = \frac{0-2}{1.2} \approx -1.67$$

and

$$z_2 = \frac{1.28-2}{1.2} = -0.6$$

So P$\{0 \leq X \leq 1.28\} \approx$ P$\{-1.67 \leq Z \leq -0.6\} \approx \Phi(1.67) - \Phi(0.6) \approx$ $0.4525 - 0.2257 \approx 0.2268$.

- Find P$\{-6.\ 20 \leq X \leq -4.52\}$ with $\mu = -5$ and $\sigma = 0.2$. Then

$$z_1 = \frac{-6.20-(-5)}{0.2} \approx \frac{-1.2}{0.2} \approx -6$$

and

$$z_2 = \frac{-4.52-(-5)}{0.2} = \frac{0.48}{0.2} = 2.4$$

This means P$\{-6.\ 20 \leq X \leq -4.52\} \approx$ P$\{-6 \leq Z \leq 2.4\} \approx \Phi(6) + \Phi(2.4) \approx 5.000 + 0.4918 \approx 0.9918$.

9.4 The Normal Approximation of the Binomial Distribution

Recall from Chapter 7 that a Bernoulli trial has two different outcomes, success with probability p and failure with probability $1-p$. The random variable B associated with Bernoulli trials counts the number k of successes in n trials. We saw that the probability of k successes in n trials is

$$\mathrm{P}\{B = k\} = {}_n\mathrm{C}_k\, p^k\, (1-p)^{(n-k)}$$

As n gets larger, this number gets more difficult to compute. The factor ${}_n\mathrm{C}_k$, the number of combinations of n objects taken k at a time, gets very large and the powers p^k and $(1-p)^{(n-k)}$ get very small. If a probability over a range, such as $P\{k \leq B \leq l\}$, is to be computed, the computational difficulty is even greater.

The normal distribution with mean $\mu = np$ and standard deviation $\sigma = \sqrt{np(1-p)}$ provides a very good approximation for the binomial distribution when n and p have the properties that $np \geq 5$ and $n(1-p) \geq 5$.

Justification for using the normal distribution as an approximation for the binomial distribution is shown in Figure 9.18, where a histogram of the binomial distribution for $n = 15$ and $p = 0.6$ is shown along with a graph of the normal distribution with mean $\mu = 15 \times 0.6 = 9$ and standard deviation $\sigma = \sqrt{15 \times 0.6 \times 0.4} \approx 1.9$.

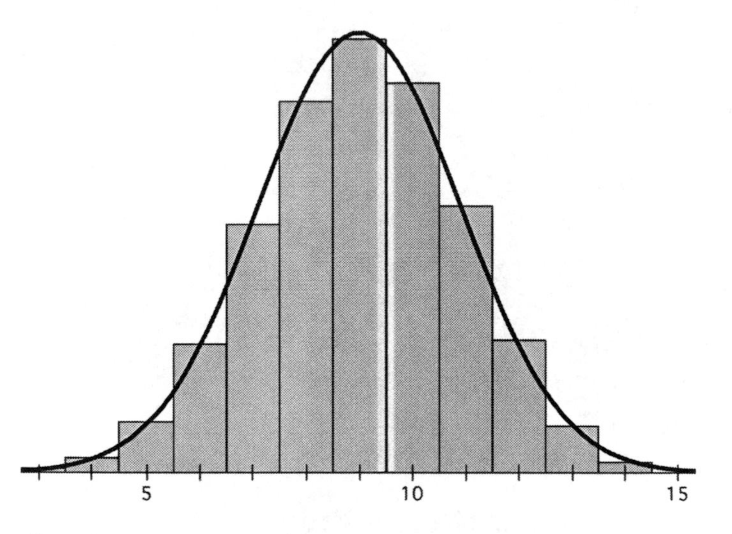

Figure 9.18
Binomial distribution with $n = 15$ trials and probability of success $p = 0.6$ graphed along with a normal distribution function with standard deviation $\sigma = 1.9$ and mean $\mu = 9$.

In this figure, we see that the curve of the normal distribution is close to the center of the columns of the histogram of the binomial distribution.

The formula giving the normal distribution approximation of the binomial distribution is

$$ {}_nC_k\, p^k (1-p)^{(n-k)} = P\{B = k\} \approx P\{k - 0.5 \leq X \leq k + 0.5\} $$

where X is the normal random variable with mean $\mu = np$ and standard deviation $\sigma = \sqrt{np(1-p)}$ and B is the binomial random variable, which gives the number of successes in n trials of a binomial experiment with probability of success equal to p.

If you want to find the probability that the number of successes of the binomial experiment lies in an interval, say from k_1 to k_2, this formula becomes

$$P\{ k_1 \le B \le k_2\} \approx P\{k_1 - 0.5 \le X \le k_2 + 0.5\}$$

Another way to view this approximation is shown in Figure 9.19. A device constructed by Sir Francis Galton, a 19th-century British statistician, simulates a binomial distribution. Marbles or balls are dropped into the top of the device and then are deflected left or right at random by a pin or block. After several layers of pins, the marbles are collected in bins at the bottom. Since left and right paths are equally likely at each stage, the heights of the columns of marbles at the bottom show a binomial distribution. These heights are a good approximation to the normal distribution, as shown in the picture.

Figure 9.19
Device to show the normal approximation to the binomial distribution.

Examples of the Normal Approximation of the Binomial Distribution

- A fair coin is tossed 50 times. Find the probability that heads appear exactly 30 times.

  Solution: This is a binomial distribution with $p = 0.5$, $n = 50$, and $k = 30$. Since $np = 50 \times 0.5 = 25 \geq 5$ and $n(1 - p) = 50 \times 0.5 = 25 \geq 5$, we can approximate this binomial distribution with the normal distribution having mean $\mu = np = 25$ and $\sigma = \sqrt{np(1-p)} = \sqrt{12.5} \approx 3.5355$. Thus

  $$_{50}C_{30}\,(0.5)^{30}\,(0.5)^{20} = P\{B = k\} \approx P\{29.5 \leq X \leq 30.5\}$$

  where X is a normal distribution with $\mu = 25$ and $\sigma = 3.5355$. To compute this using the table for the standard normal distribution, let $z_1 = (29.5 - 25) / 3.5355 \approx 1.2728$ and let $z_2 = (30.5 - 25) / 3.5355 \approx 1.5556$. Then we use the table in Figure 9.7 to calculate $P\{1.2788 \leq Z \leq 1.5556\} \approx \Phi(1.5556) - \Phi(1.2788) \approx 0.4401 - 0.3995 \approx 0.0406$. Direct computer calculation of $_{50}C_{30}\,(0.5)^{30}\,(0.5)^{20}$ gives approximately 0.04186. The error in this case is 0.00126 / 0.04186 or a little more than 3%.

- A fair coin is tossed 60 times. Find the probability that heads appear at least 26 times but no more than 32 times.

  Solution: This is a binomial distribution with $p = 0.5$, $n = 60$, and k ranging from 26 to 32. Since $np = 60 \times 0.5 = 30 \geq 5$ and $n(1 - p) = 60 \times 0.5 = 30 \geq 5$, we can use the normal distribution with mean $\mu = np = 30$ and $\sigma = \sqrt{np(1-p)} = \sqrt{15} \approx 3.8730$.

  Thus, $P\{26 \leq B \leq 32\} \approx P\{25.5 \leq X \leq 32.5\}$ where X is a normal distribution with $\mu = 30$ and $\sigma = 3.8730$. To compute this using the table for the standard normal distribution, let $z_1 = (25.5 - 30) / 3.8730 \approx -1.162$ and let $z_2 = (32.5 - 30) / 3.8730 \approx 0.645$. Then we use the table in Figure 9.7 to calculate $P\{-1.162 \leq Z \leq 0.645\} \approx \Phi(0.645) + \Phi(1.162) \approx 0.2406 + 0.3774 \approx 0.6180$. Direct computer calculation gives an answer of 0.6180, so in this case the approximation is accurate to four decimal places.

Chapter 9 Summary

The probability density function of the normal distribution is given by the equation

$$f(x) = \frac{1}{\sqrt{2\pi}\sigma} e^{-(x-\mu)^2/2\sigma^2}$$

where μ and σ are parameters.

The normal distribution with $\mu = 0$ and $\sigma = 1$ is the **standard normal distribution.**

Important features of the normal distribution:

- The graph is symmetrical about its center.
- The graph tapers off gradually to values closer and closer to 0 on both sides.
- The area between the graph and the x-axis is equal to 1.
- The overall shape of the graph is like a bell.

The parameter μ is the **mean** of the distribution. Half of the area under the graph is to the right of the mean, and half of the area under the graph is to the left of the mean. The probability of getting a value below the mean is 0.5, and the probability of getting a value above the mean is also 0.5. The highest point of the graph is directly above the mean.

For the standard normal distribution, the mean is 0 and the value of the distribution function at the mean is approximately 0.39894.

The parameter σ is the **standard deviation**, which tells how spread out the distribution is. The square of the standard deviation is called the **variance** and is denoted σ^2.

An **inflection point** is a point where concavity of a graph changes, either from concave down to concave up or from concave up to concave down.

The graph of a normal distribution is concave up on the left, then concave down, and then concave up again. There are two inflection points, one on either side of the mean μ and located at a distance equal to the standard deviation σ away from the mean.

- Approximately 68.3% of the total area under the curve is between $\mu - \sigma$ and $\mu + \sigma$.
- Approximately 95.4% of the total area is between $\mu - 2\sigma$ and $\mu + 2\sigma$.
- Approximately 99.7% of the total area is between $\mu - 3\sigma$ and $\mu + 3\sigma$.

To estimate the standard deviation σ from the graph of a normal distribution, either locate one of the inflection points and estimate its distance from the mean or estimate the maximum value that the function takes on, set that number equal to $0.399 / \sigma$ and solve for σ.

The random variable corresponding to the standard normal distribution is usually denoted by Z, while the random variable corresponding to any other normal distribution is denoted by X or some letter appropriate to the context.

The probability that the random variable Z takes on a value between a and b, in the interval $[a, b]$, is written $P\{a \le Z \le b\}$ and corresponds to the area underneath the graph of the distribution function over the interval $[a, b]$.

The area under the standard normal distribution graph from $x = 0$ to $x = z$ is called $\Phi(z)$, so $\Phi(z) = P\{0 \le X \le z\}$.

NOTE

Computing Probabilities for the Standard Normal Distribution

When computing a probability for a standard normal distribution, draw the graph and mark the interval desired.

- $P\{0 \le Z \le a\} = \Phi(a)$
- $P\{-a \le Z \le 0\} = \Phi(a)$
- $P\{-a \le Z \le b\} = \Phi(a) + \Phi(b)$
- $P\{a \le Z \le b\} = \Phi(b) - \Phi(a)$
- $P\{-a \le Z \le -b\} = \Phi(a) - \Phi(b)$
- $P\{a \le Z\} = 0.5 - \Phi(a)$
- $P\{Z \le -a\} = 0.5 - \Phi(a)$
- $P\{Z \le a\} = 0.5 + \Phi(a)$
- $P\{-a \le Z\} = 0.5 + \Phi(a)$

If X is normal distribution with mean μ and standard deviation σ, it corresponds to the standard normal distribution Z using the formula

$$z = \frac{x - \mu}{\sigma}$$

or

$$F(x) = \Phi\left(\frac{x - \mu}{\sigma}\right)$$

For the random variable X having normal distribution with mean μ and standard deviation σ,

$$P\{a \le X \le b\} = P\{z_1 \le Z \le z_2\}$$

where

$$z_1 = \frac{a - \mu}{\sigma} \text{ and } z_2 = \frac{b - \mu}{\sigma}$$

Linear interpolation is a technique for estimating a number from other numbers in a table. If the numbers p and q correspond to numbers a and b in a table, and x is between a and b, then the number corresponding to x is approximately equal to

$$(q - p)\left(\frac{x - a}{b - a}\right) + p$$

The random variable B associated with a Bernoulli experiment giving the number k of successes in n trials, with probability p of success and probability of $(1 - p)$ of failure, can be approximated by

$$P\{B = k\} = {}_nC_k\, p^k (1 - p)^{(n-k)} \approx P\{k - 0.5 \le X \le k + 0.5\}$$

if n and p have the properties that $np \ge 5$ and $n(1 - p) \ge 5$ and X is the normal random variable with mean $\mu = np$ and standard deviation

$\sigma = \sqrt{np(1 - p)}$.

For the same parameters, the probability that the number of successes of the binomial experiment lies in the interval from k_1 to k_2 is

$$P\{k_1 \le B \le k_2\} \approx P\{k_1 - 0.5 \le X \le k_2 + 0.5\}$$

Chapter 9 Practice Problems

1. Find the mean μ and standard deviation σ of the following distribution functions.

   a. $f(x) = \dfrac{1}{\sqrt{2\pi}2} e^{-(x+4)^2/8}$

   b. $f(x) = \dfrac{1}{0.5\sqrt{2\pi}} e^{-x^2/\,0.5}$

   c. $f(x) = \dfrac{1}{\sqrt{2\pi}3} e^{-(x-4)^2/18}$

2. Give the normal distribution functions that have the following means and standard deviations.

   a. $\mu = 3$ and $\sigma = 0.5$

   b. $\mu = 2$ and $\sigma = 1$

   c. $\mu = -1$ and $\sigma = 4$

3. For the three graphs in Figure 9.20, give the approximate mean μ and standard deviation σ.

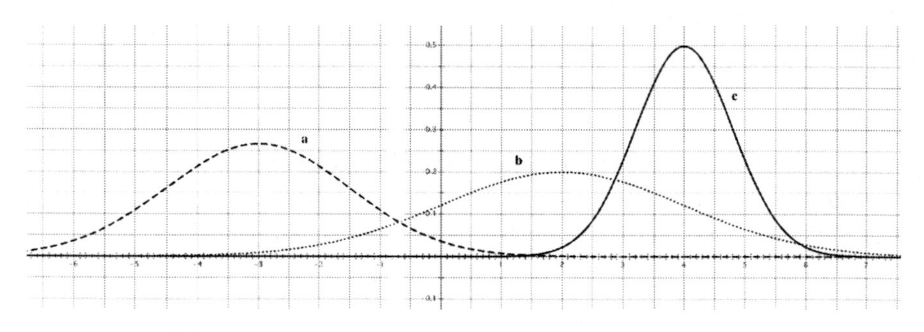

Figure 9.20
Three normal distributions.

4. Sketch the graphs of normal distributions with the following statistics.

   a. $\mu = -2$ and $\sigma = 3$

   b. $\mu = 1$ and $\sigma = 0.5$

   c. $\mu = 3$ and $\sigma = 2$

5. A normal distribution has height 0.285 at its mean μ. What is its standard deviation σ?

6. A normal distribution has height 1.33 at its mean μ. What is its standard deviation σ?

7. Find the following probabilities for the standard normal distribution Z.

a. $P\{0 \leq Z \leq 1.2\}$

b. $P\{0 \leq Z \leq 2.03\}$

c. $P\{-3.13 \leq Z \leq 0\}$

d. $P\{Z \leq 1.2\}$

e. $P\{Z \leq -0.3\}$

f. $P\{0.8 \leq Z\}$

g. $P\{-1.8 \leq Z\}$

h. $P\{0.5 \leq Z \leq 1.4\}$

i. $P\{-1.6 \leq Z \leq 0.2\}$

j. $P\{-1.3 \leq Z \leq -0.6\}$

8. Use linear approximation to find the following values.

a. $\Phi(0.182)$

b. $\Phi(1.005)$

c. $\Phi(1.551)$

d. $\Phi(2.338)$

9. Find the following probabilities for the normal distributions X with the given means μ and standard deviations σ.

a. $P\{0 \leq X \leq 1.2\}$ with $\mu = 3$ and $\sigma = 4$

b. $P\{0 \leq X \leq 2.03\}$ with $\mu = -1$ and $\sigma = 2.5$

c. $P\{-3.13 \leq X \leq 0\}$ with $\mu = 2.8$ and $\sigma = 3$

d. $P\{-3.13 \leq X \leq 0\}$ with $\mu = -5$ and $\sigma = 1.8$

e. $P\{X \leq 3.2\}$ with $\mu = 5.8$ and $\sigma = 3.5$

f. $P\{X \leq -0.3\}$ with $\mu = -6.5$ and $\sigma = 4.5$

g. $P\{0.8 \leq X\}$ with $\mu = 0.2$ and $\sigma = 1.5$

h. $P\{-1.8 \leq X\}$ with $\mu = -3$ and $\sigma = 4.6$

i. $P\{0.5 \leq X \leq 1.4\}$ with $\mu = 5.5$ and $\sigma = 7$

j. $P\{-1.6 \leq X \leq 0.2\}$ with $\mu = 3.2$ and $\sigma = 1$

k. $P\{-8.3 \leq X \leq -0.6\}$ with $\mu = 0$ and $\sigma = 3.2$

l. $P\{-1.3 \leq X \leq -0.6\}$ with $\mu = -4$ and $\sigma = 8.5$

10. Use the normal distribution to approximate the binomial probability

$$_{20}C_6\,(0.6)^6\,(0.4)^{14}$$

11. Use the normal distribution to approximate the binomial probability

$$_{30}C_{10}\ (0.7)^{10}\ (0.3)^{20}$$

12. A die is tossed 40 times. Find the probability that a 6 appears at least 10 times but no more than 15 times.

Answers to Chapter 9 Practice Problems

1. a. $\mu = -4$ and $\sigma = 2$

b. $\mu = 0$ and $\sigma = 0.5$

c. $\mu = 4$ and $\sigma = 3$

2. a. $f(x) = \dfrac{1}{0.5\sqrt{2\pi}} e^{-(x-3)^2/0.5}$

b. $f(x) = \dfrac{1}{\sqrt{2\pi}} e^{-(x-2)^2/2}$

c. $f(x) = \dfrac{1}{\sqrt{2\pi}4} e^{-(x+1)^2/32}$

3. a. $\mu = -3$ and $\sigma = 1.5$

b. $\mu = 2$ and $\sigma = 2$

c. $\mu = 4$ and $\sigma = 0.8$

4. See Figure 9.21.

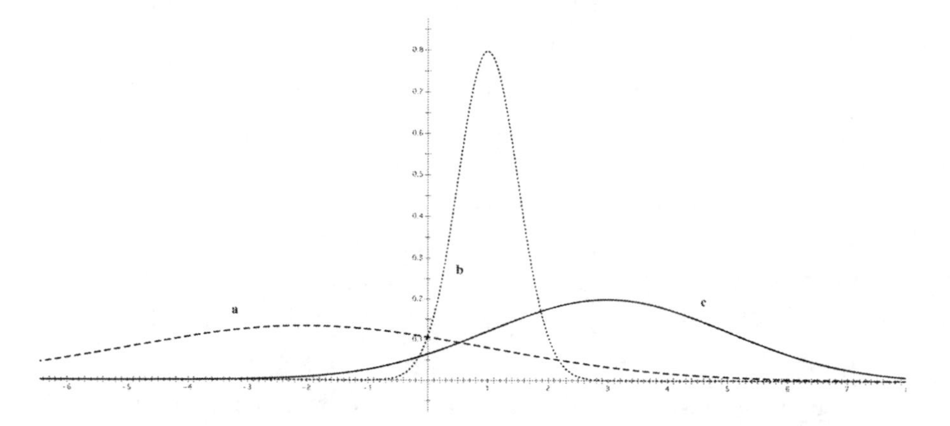

Figure 9.21
Three normal distributions.

5. $\sigma = 0.399 / 0.285 = 1.4$

6. $\sigma = 0.399 / 1.33 = 0.3$

7. **a.** $P\{0 \le Z \le 1.2\} = \Phi(1.2) \approx 0.3849$

b. $P\{0 \le Z \le 2.03\} = \Phi(2.03) \approx 0.4788$

c. $P\{-3.13 \le Z \le 0\} = \Phi(3.13) \approx 0.4991$

d. $P\{Z \le 1.2\} = 0.5 + \Phi(1.2) \approx 0.5 + 0.3849 = 0.8849$

e. $P\{Z \le -0.3\} = 0.5 - \Phi(0.3) \approx 0.5 - 0.1179 = 0.3821$

f. $P\{0.8 \le Z\} = 0.5 - \Phi(0.8) \approx 0.5 - 0.2881 = 0.2119$

g. $P\{-1.8 \le Z\} = 0.5 + \Phi(1.8) \approx 0.5 + 0.4641 = 0.9641$

h. $P\{0.5 \le Z \le 1.4\} = \Phi(1.4) - \Phi(0.5) \approx 0.4192 - 0.1915 = 0.2277$

i. $P\{-1.6 \le Z \le 0.2\} = \Phi(1.6) + \Phi(0.2) \approx 0.4452 + 0.0793 = 0.5245$

j. $P\{-1.3 \le Z \le -0.6\} = \Phi(1.3) - \Phi(0.6) \approx 0.4032 - 0.2257 = 0.1775$

8. **a.** $$\Phi(0.182) \approx (\Phi(0.19) - \Phi(0.18))\left(\frac{0.182 - 0.18}{0.19 - 0.18}\right) + \Phi(0.18) \approx 0.07218$$

b. $$\Phi(1.005) \approx (\Phi(1.01) - \Phi(1.00))\left(\frac{1.005 - 1.00}{1.01 - 1.00}\right) + \Phi(1.00) \approx 0.3426$$

c. $$\Phi(1.551) \approx (\Phi(1.56) - \Phi(1.55))\left(\frac{1.551 - 1.55}{1.56 - 1.55}\right) + \Phi(1.55) \approx 0.4395$$

d. $$\Phi(2.338) \approx (\Phi(2.34) - \Phi(2.33))\left(\frac{2.338 - 2.33}{2.34 - 2.33}\right) + \Phi(1.90) \approx 0.4903$$

9. For many of these, you will need to use interpolation. A computer algebra program will give more accurate (and slightly different) answers.

a. $P\{0 \le X \le 1.2\} = P\{-0.75 \le Z \le -0.45\} \approx \Phi(0.75) - \Phi(0.45) \approx 0.2734 - 0.1736 \approx 0.0998$

b. $P\{0 \le X \le 2.03\} = P\{0.4 \le Z \le 1.212\} \approx \Phi(1.212) - \Phi(0.4) \approx 0.3873 - 0.1554 \approx 0.2319$

c. $P\{-3.13 \le X \le 0\} = P\{-1.977 \le Z \le -0.933\} \approx \Phi(1.977) - \Phi(0.933) \approx 0.4760 - 0.3246 \approx 0.1514$

d. $P\{-3.13 \le X \le 0\} = P\{1.039 \le Z \le 2.778\} \approx \Phi(2.778) - \Phi(1.039) \approx 0.4973 - 0.3505 \approx 0.1468$

e. $P\{X \leq 3.2\} = P\{Z \leq -0.743\} \approx 0.5 - \Phi(0.743) \approx 0.5 - 0.2713 \approx 0.2287$

f. $P\{X \leq -0.3\} = P\{Z \leq 1.378\} \approx 0.5 + \Phi(1.278) \approx 0.5 + 0.4159 \approx 0.9159$

g. $P\{0.8 \leq X\} = P\{0.4 \leq Z\} \approx 0.5 - \Phi(0.4) \approx 0.5 - 0.1554 \approx 0.3446$

h. $P\{-1.8 \leq X\} = P\{0.261 \leq Z\} \approx 0.5 - \Phi(0.261) \approx 0.5 - 0.1030 \approx 0.3970$

i. $P\{0.5 \leq X \leq 1.4\} = P\{-0.714 \leq Z \leq -0.586\} \approx \Phi(0.714) - \Phi(0.586) \approx 0.2623 - 0.2210 \approx 0.0413$

j. $P\{-1.6 \leq X \leq 0.2\} = P\{-4.8 \leq Z \leq -3.0\} \approx \Phi(4.8) - \Phi(3.0) \approx 0.5000 - 0.4987 \approx 0.0013$

k. $P\{-1.3 \leq X \leq -0.6\} = P\{-2.594 \leq Z \leq -0.1875\} \approx \Phi(2.594) - \Phi(0.1875) \approx 0.4952 - 0.0743 \approx 0.4209$

l. $P\{-1.3 \leq X \leq -0.6\} = P\{0.318 \leq Z \leq 0.4\} \approx \Phi(0.4) - \Phi(0.318) \approx 0.1554 + 0.1247 \approx 0.0307$

10. The probability ${}_{20}C_6\,(0.6)^6\,(0.4)^{14}$ can be approximated by the normal distribution with $\mu = 20 \times 0.6 = 12$ and $\sigma = \sqrt{20(0.6)(0.4)} \approx 2.1909$.

$$
\begin{aligned}
{}_{20}C_6\,(0.6)^6\,(0.4)^{14} &\approx P\{6 - 0.5 \leq X \leq 6 + 0.5\} \\
&\approx P\{5.5 \leq X \leq 6.5\} \\
&\approx P\left\{\frac{5.5 - 12}{2.1909} \leq Z \leq \frac{6.5 - 12}{2.1909}\right\} \\
&\approx P\{-2.9668 \leq Z \leq -2.5104\} \\
&\approx \Phi(2.9668) - \Phi(2.5104) \\
&\approx 0.4985 - 0.4941 \\
&\approx 0.0044
\end{aligned}
$$

A calculator or computer algebra computation gives 0.004854, so this estimate has an error of around 9%.

11. The probability ${}_{30}C_{10}\,(0.7)^{10}\,(0.3)^{20}$ can be approximated by the normal distribution with $\mu = 30 \times 0.7 = 21$ and $\sigma = \sqrt{30(0.7)(0.3)} \approx 2.5100$.

$$
\begin{aligned}
{}_{30}C_{10}\,(0.7)^{10}\,(0.3)^{20} &\approx P\{10-0.5 \le X \le 10+0.5\} \\
&\approx P\{9.5 \le X \le 11.5\} \\
&\approx P\left\{\frac{9.5-21}{2.5100} \le Z \le \frac{10.5-21}{2.5100}\right\} \\
&\approx P\{-4.582 \le Z \le -4.183\} \\
&\approx \Phi(4.582) - \Phi(4.183) \\
&\approx 0.5000 - 0.500 \\
&\approx 0.0000
\end{aligned}
$$

In a Bernoulli experiment with 30 trials and probability of success equal to 0.7, the probability of exactly 10 successes can't be zero, but this normal approximation only gives an answer correct to four decimal places. A computer calculation gives an answer of 0.00002959, so the approximation turns out to be accurate to four decimal places.

12. This is a binomial distribution with $p = 1/6 \approx 0.1667$, $n = 40$, and k ranging from 10 to 15. Since $np \approx 40 \times 0.1667 \approx 6.668 \ge 5$ and $n(1-p) \approx 40 \times 0.8333 \approx 33.332 \ge 5$, we can approximate the binomial distribution using the normal distribution with mean $\mu = np = 6.668$ and $\sigma = \sqrt{np(1-p)} \approx \sqrt{5.556} \approx 2.3572$. Thus, $P\{10 \le B \le 15\} \approx P\{9.5 \le X \le 15.5\}$ where X is a normal distribution with $\mu = 6.668$ and $\sigma = 2.3572$. To compute this using the table for the standard normal distribution, let

$$z_1 = (9.5 - 6.668) / 2.3572 \approx 1.201$$

and let

$$z_2 = (15.5 - 6.668) / 2.3572 \approx 3.747$$

Then we use the table in Figure 9.7 to calculate

$$
\begin{aligned}
P\{1.201 \le Z \le 4.171\} &\approx \Phi(3.747) - \Phi(1.201) \approx 0.4999 \\
&- 0.3851 \approx 0.1148
\end{aligned}
$$

Direct computer calculation gives an answer of 0.1171, so in this case the approximation has an error 0.0023 / 0.1171, or less than 2%.

Chapter 10

Expected Value

The probability of an event or outcome tells us how likely that outcome will be when we make a random trial or experiment. A random variable allows us to focus attention on something that depends on the outcome of a random trial but may be more important to us—for example, the amount we would earn or win if a certain event happens.

In this chapter, we will study expected value or expectation, a number that depends on a random variable associated with a random trial or experiment. Expected value gives information that can be used to evaluate a random situation or compare different random situations.

A good way to think of expected value is to look at the random variable as a **payoff**, a sum of money that you receive (or pay out) when a certain event occurs. The expected value tells you how much you can expect to win or lose *in the long run*.

Another way is to view expected value as a weighted average or the center of gravity of the distribution of the random variable.

Expected value is useful because it gives us just one number that represents all of the data in a random experiment. Knowing the expected value, we can often make a reasonable judgment about a random event.

10.1 Expected Value

Suppose that you can play a game where you roll one die and receive as many dollars as the number on the die. If you roll the die once, you have an equal chance of winning \$1, \$2, \$3, \$4, \$5, or \$6. Since the probability of each number turning up is 1/6, you will win \$1 one-sixth of the time, \$2 one-sixth of the time, and so on.

For this game, the expected value is equal to

$$\left(1 \times \frac{1}{6}\right) + \left(2 \times \frac{1}{6}\right) + \left(3 \times \frac{1}{6}\right) + \left(4 \times \frac{1}{6}\right) + \left(5 \times \frac{1}{6}\right) + \left(6 \times \frac{1}{6}\right) = \frac{21}{6} = 3.5$$

This is the sum of the products of each value of the random variable multiplied by its probability.

Of course, since how much you win on each play of the game is random, you cannot predict how much you will win on any given roll. But this number gives you a kind of average that is very useful for evaluating a random outcome. In the long run, you could expect to win, *on average*, \$3.50 each time you played the game.

In this particular game, you would never actually win \$3.50, since the payoff is \$1, \$2, \$3, \$4, \$5, or \$6. In fact, in most cases, the expected value will be a value that will never occur or will occur only rarely. The expected value tells you something different—it is a number that tells you what the outcome will be *on average* after a large number of trials. Using the expected value, you can evaluate a random process. For example, the random process we see here has an expected value of \$3.50. If this were a game with the random variable giving the winnings, it would be advantageous to play if the cost to play the game were less than \$3.50, and it would be disadvantageous to play if the cost were more than \$3.50. If the cost were \$3.50, you would expect, in the long run, not to win or lose anything.

To compute the **expected value** for a random variable, you must know the distribution of the random variable. Suppose the random variable X has n different values, $x_1, x_2, x_3, \ldots, x_n$, and that the probability distribution is given by $f(x)$. This means that the probability of the value x_i happening is equal to $f(x)$. The expected value for the random variable X is denoted $E[X]$ and is computed using the following formula:

$$E[X] = x_1 f(x_1) + x_2 f(x_2) + x_3\, f(x_3) + \ldots + x_n f(x_n)$$

Using summation notation, this can be expressed as

$$\sum_{i=1}^{n} x_i f(x_i)$$

You can think of the expected value as a weighted average, where the weight assigned to each value of the random variable is the probability of that value. The expected value of the random variable associated with a random experiment is also called the **expectation, long-run value**, or **mean**.

Comparing the expected values of several different random variables gives you a way to compare the different experiments that give the random variables. This is because the expected value is one number that takes into account all of the data involved in each of the distributions and gives one number to summarize or represent the distribution.

For example, an oil company may have to decide between two different investments. Suppose drilling in location A at a cost of $10 million has a 50% chance of locating a well that will earn $50 million dollars, a 30% chance of locating a well that will earn $100 million dollars, and a 20% chance of finding nothing. Also suppose drilling in location B at a cost of $12 million has a 40% chance of locating a well that will earn $150 million dollars and a 60% chance of finding nothing. The expected value for location A is $\$50{,}000{,}000 \times 0.5 + \$100{,}000{,}000 \times 0.3 + \$0 \times 0.2 = \$55{,}000{,}000$ but the cost is $10 million, so net earnings are $45,000,000. For location B, the expected value is $\$150{,}000{,}000 \times 0.4 + \$0 \times 0.6 = \$60{,}000{,}000$ with net earnings of $48,000,000. We see that drilling at location B is a better investment.

10.2 Expected Value for Discrete Random Variables

In the following sections, we will see how to compute the expected value for some common random variables.

Coin Toss

A coin is tossed. The value of the random variable X is 1 for heads, 0 for tails. The expected value is $E[X] = 1 \times 0.5 + 0 \times 0.5 = 0.5$.

If there is a payoff of \$3 for heads and a loss of \$1 for tails, we use another random variable Y with value 3 for heads and -1 for tails. The expected value for this is $E[Y] = 3 \times 0.5 + (-1) \times 0.5 = 1$. In the long run, you could expect to win, on average, \$1 each time you played this game. If you paid \$1 to play this game, you would not win or lose anything in the long run. If instead you paid \$2 to play the game, you would lose on average \$1 − \$2 = \$1 per game.

Cards

Draw a card from a standard deck of 52. Let X be the random variable that gives the number of the card, counting the ace as 1 and face cards as 10. The expected value can be computed using the data in this table, where x is the number on the card and $f(x)$ gives the probability distribution:

x	1	2	3	4	5	6	7	8	9	10
$f(x)$	1 / 13	1 / 13	1 / 13	1 / 13	1 / 13	1 / 13	1 / 13	1 / 13	1 / 13	4 / 13

From the formula

$$E(X) = \sum_{i=1}^{n} x_i f(x_i)$$

we get

$$\begin{aligned} E(X) &= 1/13 + 2/13 + 3/13 + 4/13 + 5/13 \\ &\quad + 6/13 + 7/13 + 8/13 + 9/13 + 10 \times (4/13) \\ &= 85/13 \\ &\approx 6.538 \end{aligned}$$

This makes sense because the higher number of face cards makes the expected value larger than the middle value of 5.5.

Lottery

A lottery with a payoff of \$20 million dollars has advertised odds of winning equal to 1 in 100 million. A ticket costs \$1. Let X be the random variable that gives the net winnings a player gets by playing the lottery. The random variable takes on value $2 \times 10^7 - 1$ (this is the payoff minus the cost of the ticket) with probability 1×10^{-8} when you win, and value -1 (this is the cost of the ticket) with probability $1 - 1 \times 10^{-8} = 0.99999999$ when you lose. Thus, the expected value is

$$E[X] = (2 \times 10^7 - 1)(1 \times 10^{-8}) + (-1)(1 - 1 \times 10^{-8})$$
$$= 1{,}999{,}999 \times 0.00000001 + (-1) \times 0.99999999$$
$$= -0.8$$

This says that you will lose, on average, \$0.80 each time you play this lottery. A lottery with this payoff and odds is not a good investment.

This example highlights how expected value can be used. Expected value does not necessarily give a value that you could actually attain; in fact, most of the time it does not. However, expected value gives a number that can be used to take into account the probabilities of all of the different outcomes and gives a number that is readily understood and easily applied.

Bernoulli Trial

A Bernoulli trial has probability of success p and probability of failure $(1 - p)$. If the random variable X gives 1 for success and 0 for failure, then the expected value is

$$E[X] = 1 \times p + 0 \times (1 - p) = p$$

Now suppose that the random variable X gives a payoff for success equal to W (for winnings) and a payoff for failure equal to L (for losses). The expected value of the Bernoulli trial is $E[X] = W \times p + L \times (1 - p)$.

Examples of Expected Value for Bernoulli Trials

- An urn contains four yellow marbles and eight green marbles. You want to choose a green marble. The expected value, if green is counted as a success, is $1 \times (8 / 12) + 0 \times (4 / 14) = 2 / 3$.
- Flip a coin. If you win \$4 when heads comes up and lose \$2 when tails comes up, the expected value is \$1 since $4 \times 0.5 + (-2) \times 0.5 = 1$.

Binomial Experiment

A binomial experiment with repeated trials has probability of success p and probability of failure is $(1 - p)$ for each trial. The random variable X gives the number of successes in n trials. Computing the expected value for this binomial experiment involves several steps. The probability of i successes is

$$_nC_i\, p^i (1-p)^{n-i} = \frac{n!}{(n-i)!i!} p^i (1-p)^{n-i}$$

Then the expected value is

$$E[X] = \sum_{i=1}^{n} i \,_nC_i\; p^i (1-p)^{n-i} = \sum_{i=1}^{n} i \,\frac{n!}{(n-i)!i!}\, p^i\; (1-p)^{n-i}$$

To evaluate this sum, start by cancelling the factor i with the first factor of $i!$ in the denominator to get

$$E[X] = \sum_{i=1}^{n} \frac{n!}{(n-i)!(i-1)!}\; p^i\; (1-p)^{n-i}$$

In the next step, factor out n from $n!$ and p from p^i to get

$$E[X] = np \sum_{i=1}^{n} \frac{(n-1)!}{(n-i)!(i-1)!}\; p^{i-1}\; (1-p)^{n-i}$$

The sum in this expression now looks like a sum that we can evaluate using the binomial theorem given in Chapter 4. The binomial theorem says that

$$(a + b)^n = {_nC_n}\, a^n + {_nC_{n-1}} a^{n-1}b + {_nC_{n-2}} a^{n-2}b^2 + {_nC_{n-3}} a^{n-3}b^3$$
$$+ \ldots + {_nC_1} ab^{n-1} + {_nC_0} b^n$$

In summation notation, this is

$$(a+b)^n = \sum_{k=0}^{n} {_nC_k}\; a^k\; b^{(n-k)}$$

To make this formula fit the present context, replace a by p and b by $(1 - p)$, getting

$$(p + (1 - p))^{n-1} = \sum_{k=0}^{n-1} {}_{n-1}C_k \; p^k \; (1-p)^{(n-1-k)}$$

$$= \sum_{k=0}^{n-1} \frac{(n-1)!}{(n-1-k)!(k)!} \; p^k \; (1-p)^{(n-1-k)}$$

This still doesn't look like the sum in the formula for $E[X]$, but if we replace k by $i - 1$, i will range from 1 to n and we get

$$\left(p+(1-p)\right)^{n-1} = \sum_{i=1}^{n} {}_{n-1}C_{i-1} \; p^{i-1} \; (1-p)^{(n-i)} = \sum_{i=1}^{n} \frac{(n-1)!}{(n-i)!(i-1)!} \; p^{i-1} \; (1-p)^{(n-i)}$$

Now we see that

$$E[X] = np \sum_{i=1}^{n} \frac{(n-1)!}{(n-i)!(1-i)!} \; p^{i-1} \; (1-p)^{n-i} = np\left(p+(1-p)\right)^{n-1} = np(1)^{n-1} = np$$

This says that for a binomial experiment with n trials and probability of success p, we can expect np successes.

Examples of Expected Values for Binomial Experiments

- Toss a coin eight times. Find the expected number of heads. There are eight trials with probability of success equal to 0.5, so the expected number of heads is $8 \times 0.5 = 4$.
- Toss a die 24 times. Find the expected number of times a 6 comes up. There are 24 trials with probability of success equal to 1 / 6, so the expected number of times 6 comes up is $24 \times (1 / 6) = 4$.

Poisson Distribution

As we saw in Chapter 7, the Poisson distribution is used for events that happen at random in an interval of time or region of space. The number of events possible can be any whole number starting with 0. The Poisson random variable giving the number k of such events in an interval of size t is

$$P\{X = k\} = e^{-\lambda t} \frac{(\lambda t)^k}{k!}$$

where λ is a constant that gives the average rate of occurrence of the random events in time or space.

To determine the expected value, we must add up all terms like

$$ke^{-\lambda t} \frac{(\lambda t)^k}{k!}$$

where k can take on any value 0, 1, 2, 3, ... This means that we have to add up an infinite number or terms, and the expected value for the Poisson distribution is

$$E[X] = \sum_{i=0}^{\infty} ie^{-\lambda t} \frac{(\lambda t)^i}{i!}$$

Techniques of calculus can be used to find sums like this and give the result $E[X] = \lambda$.

Thus the expected value of the Poisson random variable with distribution

$$P\{X = k\} = e^{-\lambda t} \frac{(\lambda t)^k}{k!}$$

is $E[X] = \lambda$.

Examples of Expected Values for the Poisson Distribution

- Suppose that the average number of misprints on a page in a book of 600 pages is 0.667; this is a Poisson distribution with parameter λ = 0.667. The expected number of misprints on a given page is thus the same as λ and is 0.667.
- Suppose that calls to a help desk have a Poisson distribution with λ = 17.3 calls per hour. The expected number of calls per hour is 17.3.

Geometric Distribution

The geometric random variable gives the number of trials until the first success in a sequence of Bernoulli trials. As we saw in Chapter 7, its distribution is given by

$$P\{X = n\} = (1 - p)^{(n-1)} \times p$$

where the probability of success is p, and n is the number of the first successful trial.

As with the Poisson distribution, techniques of calculus are necessary to find the expected value of the geometric distribution. These techniques show that for the geometric random variable with distribution $P\{X = n\} = (1 - p)^{(n-1)} \times p$, the expected value is $E[X] = 1 / p$.

Examples of Expected Values for the Geometric Distribution

- Toss a coin until you get the first head. The probability of success is $p = 0.5$, and the expected number of trials is $1 / 0.5 = 2$.
- Draw a card from a standard deck of 52 until you get a jack; replace the card after a trial when you do not get a jack. The probability of success on each trial is $p = 4 / 52 = 1 / 13$, so the expected number of trials is $1 / (1 / 13) = 13$.
- Draw a card from a standard deck of 52 until you get the jack of diamonds; replace the card after a trial when you do not get the jack of diamonds. The probability of success on each trial is $p = 1 / 52$, so the expected number of trials is $1 / (1 / 52) = 52$.

Negative Binomial Distribution

If the probability of success in a binomial trial is p, then the number of trials until we achieve r successes is given by the negative binomial random variable. We recall from Chapter 7 that the general formula for the probability of getting r successes in exactly n trials is

$$P\{X = n\} = {}_{n-1}C_{r-1} \times p^r \times (1 - p)^{n-r}$$

The expectation of this distribution gives the number of trials we expect in order to get r successes and is given by the sum

$$E[X] = \sum_{n=r}^{\infty} n \; {}_{n-1}C_{r-1} \; p^r \; (1-p)^{n-r}$$

This is an infinite sum and, again, the techniques of calculus are necessary to evaluate it. The answer is r / p.

We saw that for a binomial trial with probability of success equal to p, the expected number of trials to get one success was $1 / p$ and it makes sense that the expected number of trials to get r successes is r / p.

Summarizing, the expected value of the negative binomial random variable that gives the number of trials to achieve r successes, where the probability of success on each trial equal to p, is equal to r / p.

Examples of Expected Values for the Negative Binomial Distribution

- Flip a coin. We want to get one head. The probability of success p is 0.5 and the number r of successes is 1, we find $1 / 0.5 = 2$. Thus the expected number of flips that give one head is 2.
- Flip a coin. Now we want 10 heads. The probability of success p is 0.5 and the number r of successes is 10, so we compute $10 / 0.5 = 20$. The expected number of flips is 20.
- Throw two dice. We want to get two 7s. The probability of success p is $6 / 36 = 1 / 6$ and the number r of successes is 2, and $2 / (1 / 6) = 12$.The expected number of throws for two 7s is 12.
- Throw two dice. This time, we want to get three 2s. The probability of success p is $1 / 36$ and the number r of successes is 3, and $3 / (1 / 36) = 108$. The expected number of throws to get three 2s is 108.

Hypergeometric Distribution

The hypergeometric distribution gives the number of successes when making n simultaneous choices from a collection of N objects with Np objects that are successes. This is the same as choosing n objects one at a time without replacement and ignoring the order of choosing the objects and is referred to as sampling without replacement. In Chapter 7, we saw that the probability distribution, with X equal to the number of successes, is given by

$$P\{X = k\} = \frac{{}_{Np}C_k \times {}_{N-Np}C_{n-k}}{{}_N C_n}.$$

Using the formula for expected value, we see that

$$E[X] = \sum_{i=1}^{n} i \frac{{}_{Np}C_i \times {}_{N-Np}C_{n-i}}{{}_N C_n} = \sum_{i=1}^{n} i \frac{(Np)!(N-Np)!}{i!(Np-i)!(n-i)!(N-Np-n+i)!} \times \frac{n!(N-n)!}{N!}$$

Using properties of binomial coefficients, this can be simplified to np. Thus the expected number of successful choices when choosing n objects from a collection of N objects, Np of which are successful, is equal to np.

To find p in a given problem, take the number of successful choices n and divide by the number of possible choices N.

Examples of Expected Values for the Hypergeometric Distribution

- You are dealt a hand of 13 cards from a standard deck. This is a choice of 13 cards from a collection of 52. To compute the expected number of spades, treat a spade as a success. Thus, there are $N = 52$ objects, of which 13 are successes, so $p = 13 / 52 = 1 / 4$. There are $n = 13$ choices. The expected value is equal to $13 \times (1 / 4) = 3.25$ spades in a hand of 13 cards.
- To find the expected number of face cards in a hand of $n = 13$ cards, we now treat the face cards as successes. We have $N = 52$ cards and $p = 12 / 52$. The expected number of face cards in a hand of 13 cards is $13 \times (12 / 52) = 3$ face cards in a hand of 13 cards.

10.3 Expected Value for Continuous Random Variables

A continuous random variable has infinitely many different possible values along an interval. The probability distribution is a curve, and the probability of landing in an interval is the same as the area under the distribution function above the interval. The expectation for a continuous random variable has a definition that is similar to that of a discrete random variable, but it needs calculus for its definition.

Intuitively, the expected value is the x-coordinate of the center of gravity of the area under the distribution curve. This means that if the graph is symmetric with respect to an x-value, then that value will be the mean. For other distributions, the expected value must be computed using calculus.

Uniform Distribution

For the uniform distribution, all outcomes in an interval $[a, b]$ are equally likely. The graph is symmetric about the x-value $(a + b) / 2$, so that is the expected value.

Examples of Expected Values for the Uniform Distribution

- The uniform distribution with values from -3 to 5 and height 0.125 has the expected value $(-3 + 5) / 2 = 2 / 2 = 1$.
- The uniform distribution with values from 6 to 16 and height 0.1 has the expected value $(6 + 16) / 2 = 22 / 2 = 11$.

Normal Distribution

The normal distribution has density function given by the equation

$$f(x) = \frac{1}{\sqrt{2\pi}\sigma} e^{-(x-\mu)^2/2\sigma^2}$$

where μ and σ are parameters. As we have seen, the graph of the normal distribution is symmetric about the x-value μ, which is the mean. Therefore, the expected value of the normal distribution is equal to its mean, $E[X] = \mu$.

Examples of Expected Values for the Normal Distribution

- The expected value of the normal distribution

$$f(x) = \frac{1}{\sqrt{2\pi 2}} e^{-(x+4)^2/8}$$

  is −4.

- The expected value of the normal distribution

$$f(x) = \frac{1}{\sqrt{2\pi 3}} e^{-(x-4)^2/18}$$

  is 4.

Exponential Distribution

The exponential distribution, given by the formula

$$f(t) = \lambda e^{-\lambda t} \text{ for } t \geq 0 \text{ and } \lambda > 0$$

models the time until a random event happens. The expected value for this distribution must be computed using calculus, but it turns out to be very simple: $E[X] = 1 / \lambda$.

Examples of Expected Values for the Exponential Distribution

- The expected value of the exponential distribution $f(t) = 6e^{-6t}$ is $1 / 6$.
- The expected value of the exponential distribution $f(t) = -5e^{5t}$ is $-1 / 5 = -0.2$.

10.4 Expectation of a Multiple or Sum of Random Variables

The expected value of a multiple or sum of random variables is very easy to compute. It is just the multiple or sum of the expected values of the random variables.

Let X and Y be two random variables that have the same underlying sample space, and let a and b be real numbers. Then the following formulas show how to compute multiples or sums of random variables.

$$E[aX] = aE[X]$$

$$E[X + b] = E[X] + b$$

$$E[X + Y] = E[X] + E[Y]$$

Examples of Expected Values for Sums and Multiples of Expected Values

- If two random variables X and Y have expectations $E[X] = 9$ and $E[Y] = 4$, then

  a. $E[X + Y] = E[X] + E[Y] = 9 + 4 = 13$
  b. $E[5X] = 5E[X] = 5 \times 9 = 45$
  c. $E[X + 8] = E[X] + 8 = 9 + 8 = 17$
  d. $E[X] + E[7Y] = E[X] + 7E[Y] = 9 + (7 \times 4) = 37$

- A single die is thrown and a card is drawn from a standard deck of 52. What is the sum of the number on the die and the number on the card (counting face cards as 10)? Let X give the number on the die, and let Y give the value of the card. We want $E[X + Y]$, so we find $E[X] + E[Y]$, which is easier to compute.

$$\begin{aligned} E[X] &= 1\times(1/6)+2\times(1/6)+3\times(1/6)+4\times(1/6)+5\times(1/6)+6\times(1/6) \\ &= 21/36 \\ &= 3.5 \end{aligned}$$

$$\begin{aligned} E[Y] &= 1\ (1/13)+2\times(1/13)+3\times(1/13)+4\times(1/13)+5\times(1/13)+6\times(1/13) \\ &\quad +7\times(1/13)+8\times(1/13)+9\times(1/13)+10\times(4/13) \\ &= 85/13 \\ &\approx 6.538 \end{aligned}$$

Finally, $E[X + Y] = E[X] + E[Y] \approx 3.5 + 6.538 \approx 10.038$.

10.5 Conditional Expectation

Let X and Y be two jointly distributed random variables that have the same underlying sample space. We want to compute the expected value of X given that Y has happened. Recall the definition of conditional probability from Chapter 6: P{A | B} = P(AB) / P(B). For random variables, the formula for conditional probability becomes

$$P\{X = x \mid Y = y\} = P\{X = x \text{ and } Y = y\} / P\{Y = y\}$$

The conditional expectation is therefore

$$E[X \mid Y = y] = \sum_x x\, P\{X = x \mid Y = y\} = \sum_x x \frac{P\{X = x \text{ and } Y = y\}}{P\{Y = y\}}$$

The factor $P\{Y = y\}$ can be factored out, giving an alternative formula that is usually easier to compute

$$E[X \mid Y = y] = \left(\sum_x x\ P\{X = x \text{ and } Y = y\}\right) \div P\{Y = y\}$$

This tells us what the expected value of the random variable X is once we know that the random variable Y has already taken on value y.

Example of Conditional Expectations

Throw a pair of dice. Let X give the sum of the two dice and let Y give the higher of the two dice. To find the expected value for the sum if you know the higher value is 3, first find $P\{Y = 3\}$. There are 5 out of 36 possible throws where 3 is the higher number; these are (1, 3), (3, 1), (2, 3), (3, 2) and (3, 3), so $P\{Y = 3\} = 5 / 36$. Then find $P\{X = x \text{ and } Y = 3\}$ for every value of x. These values are given in tabular form. Note that if the higher value is 3, the sum x must be 6 or less, so we only need to look at values of x which are 6 or less.

x	2	3	4	5	6
$P\{X = x \text{ and } Y = 3\}$	0	0	2 / 36	2 / 36	1 / 36

The expected value for the sum of two dice with the higher die having value 3 is computed as follows:

$$E[X \mid Y = 3] = \left(\sum_{i=2}^{6} x \text{ P}\{X = x \text{ and } Y = 3\}\right) \div \text{P}\{Y = 3\}$$
$$= \left(2 \times 0 + 3 \times 0 + 4 \times \frac{2}{36} + 5 \times \frac{2}{36} + 6 \times \frac{1}{36}\right) \div \left(\frac{6}{36}\right)$$
$$= \frac{24}{36} \div \frac{5}{36}$$
$$= \frac{24}{5}$$
$$= 4.8$$

This makes sense because there are three possible sums—4, 5, and 6—and 6 is less likely, so the expected value should be less than 5.

Chapter 10 Summary

The expected value of a random variable tells us what might be expected to happen in the long run. One can think of the expected value as a payoff, as a weighted average, or as a center of gravity, depending on the context. Expected value is also called expectation, long-run value, or mean.

The **expected value** of a discrete random variable X is computed from the distribution of the random variable. If the random variable X has n different values, $x_1, x_2, x_3, \ldots, x_n$, and the probability distribution is given by $f(x)$, then the expected value for the random variable X is denoted $E[X]$ and is

$$E[X] = x_1 f(x_1) + x_2 f(x_2) + x_3 f(x_3) + \ldots + x_n f(x_n)$$

or

$$\sum_{i=1}^{n} x_i f(x_i)$$

You can compare several different random variables by comparing their expected values.

A Bernoulli trial with probability of success p and probability of failure is $(1 - p)$ and random variable X that gives W (win) for success and L (lose) for failure has expected value

$$E[X] = W \times p + L \times (1 - p)$$

A binomial experiment with n trials, each having probability of success p and probability of failure is $(1 - p)$ and random variable X giving the number of successes in n trials has expected value $E[X] = np$.

The Poisson random variable with parameter λ giving the average rate of occurrence of some event an interval of time or space of size t has distribution

$$\mathrm{P}\{X = k\} = e^{-\lambda t} \frac{(\lambda t)^k}{k!}$$

and expected value $E[X] = \lambda$.

The geometric random variable giving the number of trials until the first success in a sequence of Bernoulli trials where the probability of success is p and n is the number of the first successful trial has distribution function

$$\mathrm{P}\{X = n\} = (1 - p)^{(n-1)} \times p$$

and expected value $E[X] = 1 / p$.

The negative binomial random variable gives the number of trials until we achieve r successes, where the probability of success p, has distribution

$$\mathrm{P}\{X = n\} = {}_{n-1}\mathrm{C}_{r-1} \times p^r \times (1 - p)^{n-r}$$

and expectation $E[X] = r / p$.

The hypergeometric distribution, or sampling without replacement, gives the number of successes when making n simultaneous choices from a collection of N objects with Np objects that are successes. The probability distribution is

$$\mathrm{P}\{X = k\} = \frac{{}_{Np}\mathrm{C}_k \times {}_{N-Np}\mathrm{C}_{n-k}}{{}_N\mathrm{C}_n}$$

and the expected value is $E[X] = np$.

Using properties of binomial coefficients, this can be simplified to np.

The expectation of a continuous random variable is the x-coordinate of the center of gravity of the area under the distribution curve. If the graph of a continuous distribution is symmetric with respect to an x-value, then that value will be the mean.

The expected value of the uniform distribution with all outcomes in an interval $[a, b]$ has an equally likely expected value $(a + b) / 2$.

The normal distribution has density function given by the equation

$$f(x) = \frac{1}{\sqrt{2\pi\sigma}} e^{-(x-\mu)^2/2\sigma^2}$$

and expected value equal to its mean, $E[X] = \mu$.

The exponential distribution given by the formula

$$f(t) = \lambda e^{-\lambda t} \text{ for } t \geq 0 \text{ and } \lambda > 0$$

has expected value $E[X] = 1 / \lambda$.

Let X and Y be two random variables that have the same underlying sample space and let a and b be real numbers. Multiples of X and Y can be computed by the following formulas:

$$E[aX] = aE[X]$$

$$E[X + b] = E[X] + b$$

$$E[X + Y] = E[X] + E[Y]$$

Chapter 10 Practice Problems

1. Throw a die. If you win \$2 when the number is even and lose \$1 when the number is odd, what is the expected value? If you pay \$1 to play the game, will you win in the long run?
2. Remove the face cards from a standard deck, leaving the cards ace through 10 of each suit. Choose a card from this smaller deck and look at the number on the card. What is the expected value?

3. A company has a choice of three marketing strategies. The first will cost \$150,000 and has a 40% chance of \$1,500,000 in profits and a 60% chance of \$500,000 in profits. The second strategy will cost \$50,000 and has a 20% chance of \$1,000,000 in profits and an 80% chance of \$600,000 in profits. The third strategy will cost \$80,000 and has a 50% chance of \$1,000,000 in profits and a 50% chance of \$800,000 in profits. Which is the best strategy?
4. An urn contains 10 balls, three white and seven red. You win \$5 if you draw a white ball and \$2 if you draw a red ball. What is the expected value of this random game? Should you pay \$2 to play the game? Should you pay \$3 to play the game?
5. An urn contains four balls numbered 2, 5, 6, and 7. You draw one ball at random. What is the expected value of the number on the ball?
6. Suppose X is a random variable with distribution given in the table below.

x	3	6	9	12
$f(x)$	0.50	0.30	0.15	0.05

Find the expected value of X.
7. Flip a coin 10 times. What is the expected number of heads?
8. Throw two dice 30 times. What is the expected number of doubles?
9. Suppose a field biologist determines that rabbit holes in a field have a Poisson distribution with $\lambda = 0.051$ rabbit holes per square meter. Choose a square meter of the field at random; what is the expected number of rabbit holes in that area?
10. Toss a die until you get 3. What is the expected number of tosses?
11. Toss two dice until you get 7. What is the expected number of tosses?
12. Flip a coin until you get 8 heads. What is the expected number of flips?
13. Throw a single die. What is the expected number of throws needed to get a 4?
14. Throw a single die. What is the expected number of throws needed to get four 6s?
15. Throw two dice. What is the expected number of throws needed to get a sum equal to 4?

16. Throw two dice. What is the expected number of throws needed to get a sum of 6 four times?
17. What is the expected number of spades in a poker hand of 5 cards taken at random from a standard deck of 52?
18. What is the expected number of face cards in a poker hand of 5 cards taken at random from a standard deck of 52?
19. What is the expected number of aces in a poker hand of 5 cards taken at random from a standard deck of 52?
20. Find the expected value of the uniform distribution with values from 2 to 18 and height 0.0625.
21. Find the expected value of the continuous random variable whose distribution is shown in Figure 10.1.

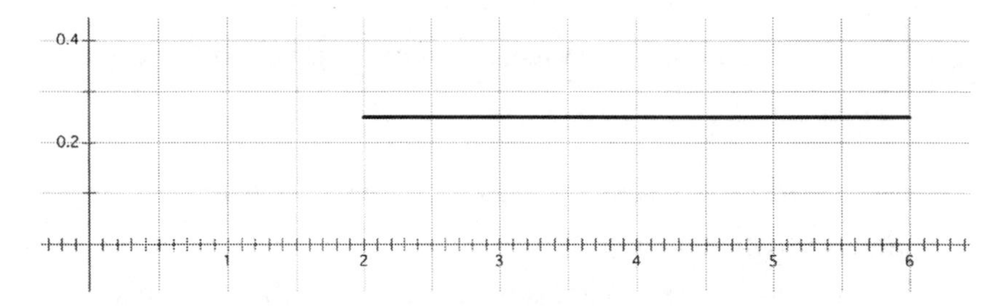

Figure 10.1
The continuous distribution for problem 21.

22. Find the expected value of the random variable whose distribution is shown in Figure 10.2.

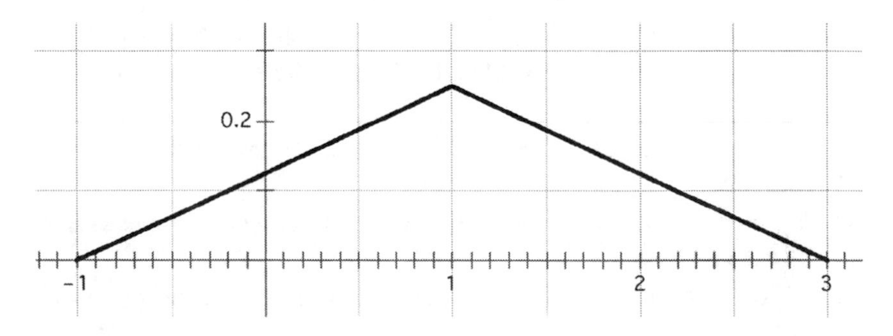

Figure 10.2
The continuous distribution for problem 22.

23. Find the expected value of the normal distribution

$$f(x) = \frac{1}{\sqrt{2\pi 2}} e^{-(x-2)^2/8}$$

24. Find the expected value of the normal distribution

$$f(x) = \frac{1}{\sqrt{2\pi 3}} e^{-(x+6)^2/18}$$

25. Find the expected value of the exponential distribution $f(t) = 6e^{-6t}$.

26. Find the expected value of the exponential distribution $f(t) = -5e^{5t}$.

27. Given two random variables X and Y and their expectations $E(X) = 3.2$ and $E(Y) = 6.4$, find the following:

a. $E(X + Y)$

b. $E(2X)$

c. $E(3X) + E(Y)$

d. $E(X + 4.6)$

28. Throw a pair of dice. Find the expected value of the higher of two dice given that the sum is 8.

29. A card is drawn from a deck of the 40 numbered cards taken from a standard deck, ace through 10. What is the expected value of the card given that it is not an ace?

Answers to Chapter 10 Practice Problems

1. The probability of each number is 1 / 6, so the expected value is

$\$2 \times (1/6 + 1/6 + 1/6) - \$1 \times (1/6 + 1/6 + 1/6) = \$2 \times 0.5 - \$1 \times 0.5 = \$0.50$

If you pay \$1 to play the game, you will lose, on average, \$0.50 per game.

2. The probability of each card is 1 / 40, but there are four cards for each number. Thus, the probability of getting each number is $4 \times (1/40) = 1/10$. The expected value is $1 \times (1/10) + 2 \times (1/10) + 3 \times (1/10) + 4 \times (1/10) + 5 \times (1/10) + 6 \times (1/10) + 7 \times (1/10) + 8 \times (1/10) + 9 \times (1/10) + 10 \times (1/10) = 55/10 = 5.5$.

3. The first strategy has expected value of profits equal to 0.4 × \$1,500,000 + 0.6 × \$500,000 = \$600,000 + \$300,000 = \$900,000. The cost is \$150,000, so the net profit is \$750,000. The second strategy has expected value of profits equal to 0.2 × \$1,000,000 + 0.8 × \$600,000 = \$200,000 + \$480,000 = \$680,000. The cost is \$50,000, so the net profit is \$630,000. The third strategy has expected value of profits equal to 0.5 × \$1,000,000 + 0.5 × \$800,000 = \$500,000 + \$400,000 = \$900,000. The cost is \$80,000, so the net profit is \$820,000. The third strategy is preferable.
4. $E(X)$ = \$5 × (0.3) + \$2 × (0.7) = \$1.50 + \$1.40 = \$2.90. If you pay \$2.00 to play the game, you will win \$0.90 on average per game, so you should play. If you pay \$3.00 to play the game, you will lose \$0.10 on average per game, so you should not play.
5. Each ball is equally likely, so the expected value is $2 \times 0.25 + 5 \times 0.25 + 6 \times 0.25 + 7 \times 0.25 = 20 \times 0.25 = 5$.
6. $E(X) = 3 \times 0.50 + 6 \times 0.30 + 9 \times 0.15 + 12 \times 0.05 = 1.50 + 1.80 + 1.35 + 0.60 = 5.25$.
7. This is a Bernoulli experiment with 10 trials and chance of success on each trial equal to 0.5. The expected number of heads is $10 \times 0.5 = 5$.
8. This is a Bernoulli experiment with 30 trials and chance of success on each trial equal to 6 / 36. The expected number of heads is $30 \times (6 / 36) = 5$.
9. The expected value is 0.051, the same as the parameter λ, since this is a Poisson distribution.
10. This is a geometric distribution. The probability of getting 3 is 1 / 6, so the expected value is $1 / (1 / 6) = 6$.
11. This is a geometric distribution. The probability of getting 7 is $6 / 36 = 1 / 6$, so the expected value is $1 / (1 / 6) = 6$.
12. This is a negative binomial distribution. The probability of success p is 0.5 and the number r of successes is 8; and $8 / 0.5 = 16$.
13. This is a negative binomial distribution. The probability of success p is 1 / 6 and the number r of successes is 1; and $1 / (1 / 6) = 6$.
14. This is a negative binomial distribution. The probability of success p is 1 / 6 and the number r of successes is 4; and $4 / (1 / 6) = 24$.
15. This is a negative binomial distribution. The probability of success p is 3 / 36 and the number r of successes is 1; and $1 / (3 / 36) = 12$.

16. This is a negative binomial distribution. The probability of success p is 5 / 36 and the number r of successes is 4; and 4 / (5 / 36) = 28.8.
17. This is a hypergeometric distribution. There are N = 52 objects, of which 52 × (13 / 52) = 52 × (1 / 4) are successes, so p = 1 / 4. There are n = 5 choices. The expected value is equal to 5 × (1 / 4) = 1.25 spades in a hand of 5 cards.
18. This is a hypergeometric distribution. There are N = 52 objects, of which 52 × (12 / 52) are successes, so p = 12 / 52. There are n = 5 choices. The expected value is equal to 5 × (12 / 52) = 1.154 face cards in a hand of 5 cards.
19. This is a hypergeometric distribution. There are N = 52 objects, of which 52 × (13 / 52) = 52 × (4 / 52) are successes, so p = 1 / 13. There are n = 5 choices. The expected value is equal to 5 × (1 / 13) = 0.385 aces in a hand of 5 cards.
20. The expected value is (2 + 18) / 2 = 20 / 2 = 10.
21. This is a continuous uniform distribution, so the expected value is (2 + 6) / 2 = 4.
22. This is a continuous distribution that is symmetric with respect to the value x = 1; therefore, the expected value is 1.
23. The expected value is the same as the mean, so it is 2.
24. The expected value is the same as the mean, so it is −6.
25. The expected value of the exponential distribution $f(t) = 6e^{-6t}$ is 1 / 6.
26. The expected value of the exponential distribution $f(t) = -5e^{5t}$ is −1 / 5 = −0.2.
27. **a.** $E(X + Y) = E(X) + E(Y) = 3.2 + 6.4 = 9.6$

    b. $E(2X) = 2E(X) = 6.4$

    c. $E(3X) + E(Y) = 3E(X) + E(Y) = 9.6 + 6.4 = 16$

    d. $E(X + 4.6) = E(X) + 4.6 = 7.8$
28. Let X be the random variable that gives the higher of the two dice and let Y be the random variable that gives the sum. $P\{Y = 8\}$ = 5 / 36. Probabilities for the higher value x are given as a table:

x	1	2	3	4	5	6
$P\{X = x \text{ and } Y = 8\}$	0	0	0	1 / 36	2 / 36	2 / 36

The expected value is

$$E[X \mid Y] = \left(1\times 0 + 2\times 0 + 3\times 0 + 4\times\left(1/36\right) + 5\times\left(2/36\right) + 6\times\left(2/36\right)\right) \div \left(5/36\right)$$

$$= \left(4/36 + 10/36 + 12/36\right) \div \left(5/36\right) = \left(26/36\right) \div \left(5/36\right) = 26/5 = 5.2$$

29. Let X be a random variable that gives the number of the card and let Y be another random variable that gives the number of the card. $P\{Y \neq 1\} = 9 / 10$. Probabilities for the value x are given as a table:

x	1	2	3	4	5	6	7	8	9	10
$P\{X = x$ and $Y \neq 1\}$	0	1 / 10	1 / 10	1 / 10	1 / 10	1 / 10	1 /10	1 / 10	1 / 10	1 / 10

The expected value is

$$E[X \mid Y \neq 1] = \left(\sum_{i=1}^{10} x \;\; E[X = x \text{ and } Y \neq 1]\right) \div P\{Y \neq 1\}$$

$$= \left(1\times 0 + \frac{2}{10} + \frac{3}{10} + \frac{4}{10} + \frac{5}{10} + \frac{6}{10} + \frac{7}{10} + \frac{8}{10} + \frac{9}{10} + \frac{10}{10}\right) \div \left(\frac{9}{10}\right) = \frac{54}{10} \div \frac{9}{10} = 6$$

This makes sense because each of the cards from 2 through 10 is equally likely once you know that you do not have an ace. Of these numbers, 6 is in the center.

Chapter 11

Laws of Large Numbers

Suppose that you flip a fair coin 100 times. This is a binomial experiment consisting of 100 independent trials. Anything, from no heads to all 100 heads, can happen, since each flip has a random outcome. We have seen that the expected number of heads is 50, and we know that the probability of getting exactly 50 heads is

$$\binom{100}{50}(0.5)^{50}(0.5)^{50} = \binom{100}{50}(0.5)^{100} \approx 0.080$$

This is quite small, but the probability of getting no heads is much, much smaller and is actually negligible. It is the same as the probability of getting all heads, and is

$$\binom{100}{100}(0.5)^{100}(0.5)^{0} = \binom{100}{0}(0.5)^{0}(0.5)^{100} \approx 1.72 \times 10^{-29}$$

The difference between these two numbers is quite striking, and our intuitive feelings about what happens in the long run are substantiated. The theorems in this chapter will give ways to describe and quantify what happens in the long run for repeated trials.

In this chapter, we will see how probability can quantify its own predictions, in four major results. The Chebyshev inequality tells the likelihood of a random variable taking on a value in an interval. The weak law of large numbers tells the likelihood of the average value of the outcomes of several trials having a value in an interval. The central limit theorem, one of the most important results in probability, uses the normal distribution to describe the long-term behavior of a sequence of many independent trials. The strong law of large numbers says that in the long run, after many trials of a random experiment, the likelihood that the average value is not the expected value is very small.

First, we look at ways that we can describe or quantify the outcome of a sequence of random trials. Then we will look at each of the results about long-term behavior.

11.1 Describing a Probability Distribution

In order to describe the long-term behavior of an experiment, we need to summarize the properties of the random variable associated with the experiment. We are already familiar with the expected value or expectation of a random variable. New tools from descriptive statistics that we will discuss in this section are the mean of a set of numbers and the variance and standard deviation of a random variable.

Recall that for a random variable X with probability distribution $f(x)$ the expected value or mean is

$$E[X] = \sum x\, f(x)$$

where the sum is taken over all possible values x of the random variable X. In the following, the expected value will often be represented by the Greek letter mu (μ). If we need to make clear that the expected value belongs to the random variable X, we write μ_X.

Examples Using Expected Values of Repeated Trials

- Recall from Chapter 10 that the expected value of a binomial experiment consisting of n trials and probability of success p is np.
- If the expected value of a random variable X is μ and if the experiment is repeated n times, then the expected value of $X + X + \cdots + X$ is $n\mu$.

The **mean** or **arithmetic mean** of a set of numbers is the sum of the numbers divided by the number of numbers. So the mean of the set {6, 12, 20, 25, 36} is

$$\frac{6 + 12 + 20 + 25 + 36}{5} = 19.8$$

A familiar use of the mean is for computing grades. If test scores are 90, 85, 92, and 98, then the mean is

$$\frac{90 + 85 + 92 + 98}{4} = 91.25$$

The mean is often called the average in everyday language, but that term is avoided in statistics because there are other kinds of averages that are used.

The expected value is like a mean and is an example of a weighted average. We can understand this by an example. Suppose the outcome of an experiment can be 10, 20, or 30. The mean of these three numbers is

$$\frac{10 + 20 + 30}{3} = 20$$

But if the probability of getting 10 is 0.2, of getting 20 is 0.3, and of getting 30 is 0.5, then the expected value of the experiment is

$$(10 \times 0.2) + (20 \times 0.3) + (30 \times 0.5) = 23$$

We see that in this case the expected value gives a better picture of the outcome of the experiment than the average does.

The variance and standard deviation give measures of how spread out the values of the random variable are or, in other words, how far the values of the random variable are from the expected value. Let's look at an example of two random experiments where the outcomes are the same: 10, 20, 30, 40, and 50. In the first experiment, with random variable *X*, the probability distribution is uniform, with distribution *f* given in the following table:

x	10	20	30	40	50
$f(x)$	0.2	0.2	0.2	0.2	0.2

The expected value for *X* is

$$\mu_X = 10 \times 0.2 + 20 \times 0.2 + 30 \times 0.2 + 40 \times 0.2 + 50 \times 0.2 = 30$$

The second experiment, with random variable *Y*, has a different probability distribution *g* shown in the next table:

y	10	20	30	40	50
$g(y)$	0.4	0.05	0.1	0.05	0.4

The expected value for *Y* is

$$\mu_Y = (10 \times 0.4) + (20 \times 0.05) + (30 \times 0.1) + (40 \times 0.05) + (50 \times 0.40 = 30$$

Both experiments have the same expected value, but the probability distributions are quite different, and the variance and standard deviation give a measure of how different the distributions are.

The **variance** of a random variable *X* is based on the difference between the expected value $E[X] = \mu$ and the values that *X* can take on. It is defined to be the expected value of the square of this difference and is denoted Var(*X*).

$$\text{Var}[X] = E\left[(X-\mu)^2\right]$$

The units of the variance are the square of the units of the random variable. So, for example, if the random variable were inches, then the variance would be inches^2. The variance will always be nonnegative; in fact, it will be positive except when there is one outcome that happens with probability 100%.

We now compute the variances for the preceding example, using a table of values. For the random variable X we get the following table:

x	10	20	30	40	50
$f(x)$	0.2	0.2	0.2	0.2	0.2
$x - \mu_X$	−20	−10	0	10	20
$(x - \mu_X)^2$	400	100	0	100	400

The variance for the random variable X is

$$E\left[\left(X-\mu_X\right)^2\right] = \sum\left(x-\mu_X\right)^2 f(x)$$

$$= 400\,(0.2) + 100\,(0.2) + 0\,(0.2) + 100\,(0.2) + 400\,(0.2) = 200$$

For the random variable Y we get another table:

y	10	20	30	40	50
$g(y)$	0.40	0.05	0.10	0.05	0.40
$y - \mu_r$	−20	−10	0	10	20
$(y - \mu_r)^2$	400	100	0	100	400

The variance for the random variable Y is

$$E\left[\left(Y-\mu_Y\right)^2\right] = \sum\left(y-\mu_Y\right)^2 g(y)$$

$$= 400\,(0.4) + 100\,(0.05) + 0\,(0.1) + 100\,(0.05) + 400\,(0.4) = 330$$

Comparing the variances, we see that the random variable Y is spread out more than the random variable X.

There is another way to compute the variance of a random variable that is usually easier. We use the properties of expected values from Chapter 10 that the expectation of a sum of random variables is the sum of the expectations of the random variables, and the expectation of a multiple of a random variable X is a multiple of the expectation of X. Suppose that we already have computed $E[X] = \mu$.

$$\begin{aligned}
\text{Var}[X] &= E\left[(X-\mu)^2\right] \\
&= E\left[X^2 - 2\mu X + \mu^2\right] \\
&= E\left[X^2\right] - E\left[2\mu X\right] + E\left[\mu^2\right] \\
&= E\left[X^2\right] - 2\mu E\left[X\right] + \mu^2 \\
&= E\left[X^2\right] - 2\mu^2 + \mu^2 \\
&= E\left[X^2\right] - \mu^2
\end{aligned}$$

For the example above, we use this method to compute the variances as follows. We create a new table for the random variable X that includes the square of the random variable, x^2:

x	10	20	30	40	50
$f(x)$	0.2	0.2	0.2	0.2	0.2
x^2	100	400	900	1600	2500

The variance for the random variable X is

$$\begin{aligned}
E\left[X^2\right] - \mu^2 &= \sum x^2 f(x) - \mu^2 \\
&= 100\,(0.2) + 400\,(0.2) + 900\,(0.2) + 1600\,(0.2) + 2500\,(0.2) - 900 \\
&= 200
\end{aligned}$$

Similarly, we get the following table for the variance of the random variable Y:

y	10	20	30	40	50
$g(y)$	0.4	0.05	0.1	0.05	0.4
y^2	100	400	900	1600	2500

The variance for the random variable Y is

$$\begin{aligned}
E\left[Y^2\right] - \mu^2 &= \sum y^2 g(y) - \mu^2 \\
&= 100\,(0.4) + 400\,(0.05) + 900\,(0.1) + 1600\,(0.05) + 2500\,(0.4) - 900 \\
&= 330
\end{aligned}$$

We get the same answers as before, but you can see that the computation is simpler.

The variance has two very important properties that we will be using later in this chapter. First, if we have a random variable X and a number a, then the variance of the random variable aX can be computed from the variance of X using the formula

$$\mathrm{Var}(aX) = a^2\,\mathrm{Var}(X)$$

Second, the variance of the sum of two random variables that are independent of one another is the sum of the variances. For two independent random variables, X and Y, we have that

$$\mathrm{Var}(X + Y) = \mathrm{Var}(X) + \mathrm{Var}(Y)$$

The variance of a random variable X is sometimes written VAR[X] or var[X].

Examples of the Variance of a Random Variable

- Let X be the random variable giving the number of successes in n trials of a binomial experiment with probability of success p. We know that each trial has value 1 with probability p and 0 with probability $(1 - p)$. The trials are independent, so the variance of X is n times the variance of one trial. Let Y be the random variable giving the number of successes in one trial. Then

$$\begin{aligned}\mathrm{Var}[Y] &= E[Y^2] - (E[Y])^2 \\ &= p - p^2 \\ &= p(1 - p)\end{aligned}$$

The variance of Y is thus $p(1 - p)$ and the variance of X is $np(1 - p)$.

- Throw a single die and let X be the number on the die. The expected value, from section 10.1 (refer to Chapter 10) is 3.5. The expected value of $E[X^2]$ is

$$\begin{aligned} E[X^2] &= 1^2\left(\frac{1}{6}\right) + 2^2\left(\frac{1}{6}\right) + 3^2\left(\frac{1}{6}\right) + 4^2\left(\frac{1}{6}\right) + 5^2\left(\frac{1}{6}\right) + 6^2\left(\frac{1}{6}\right) \\ &= \left(1 + 4 + 9 + 16 + 25 + 36\right)\left(\frac{1}{6}\right) \\ &= \frac{91}{6} \\ &\approx 15.167 \end{aligned}$$

Finally, $\text{Var}[X] = E[X^2] - (E[X])^2 \approx 15.167 - (3.5)^2 \approx 15.167 - 12.25 \approx 2.917$

Another measure of the spread of a distribution is the **standard deviation**, which is the square root of the variance. Thus for the random variable X in the preceding example, we have

$$\sigma\left(X\right) = \sqrt{\text{Var}(X)} = \sqrt{200} \approx 14.14$$

$$\sigma\left(Y\right) = \sqrt{\text{Var}(Y)} = \sqrt{330} \approx 18.17$$

The standard deviation, like the variation, is a measure of how spread out a distribution is. One advantage is that it has the same units as the random variable. So if the random variable is in inches, so is the standard deviation. Another advantage is that, being a smaller number than the variance, it may be easier to work with.

Examples of the Standard Deviation of a Random Variable

- We know the variance of a binomial experiment with n trials and probability of success p is $np(1 - p)$, so the standard deviation is $\sqrt{np(1-p)}$.
- Since the variance of the number from the toss of a single die is about 2.917, the variance is $\sqrt{2.917} \approx 1.708$.

Mathematicians like to compare the mean to the center of gravity of a physical object, the point where the mass of the object can be considered to be concentrated and still have the same properties with respect to gravity. The variation can be compared to the moment of inertia of an object, the tendency of the object to resist being rotated. And the standard distribution can be compared to the radius of gyration of the object, the radius at which all the mass can be considered to be concentrated and still have the same moment of inertia.

The mean, variation, and standard deviation can all be computed using a scientific calculator or mathematical software program.

11.2 The Chebyshev Inequality

The Chebyshev inequality gives estimates for the likelihood that a random variable will take on a value far away from its expected value.

First, we will look at a result called the Markov inequality, which gives the probability that a random variable X will take on a value larger than some number a, which is the probability $P\{X \geq a\}$. The Markov inequality is not used very much by itself, and it is limited in its scope because it will only work for a random variable that always takes on nonnegative values. However, the Markov inequality is strong enough to help us justify the Chebyshev inequality, a much more important and useful result.

To get the Markov inequality, we start with a random variable X that has only nonnegative values. This means that $x > 0$ for every value x of the random variable. Recall that the expected value is

$$E[X] = \sum_{x} x\, f(x)$$

where the sum is taken over all possible values of the random variable X. We break the sum into two pieces,

$$E[X] = \sum_{x<a} x\, f(x) + \sum_{x\geq a} x\, f(x)$$

The terms in both sums are positive because all values x are nonnegative and all values of $f(x)$ are nonnegative because $f(x)$ is a probability. If we drop the first summation, which is some nonnegative number, we get an inequality

$$E[X] \geq \sum_{x\geq a} x\, f(x)$$

For the terms in the sum in this inequality, $x \geq a$, so of course $x\, f(x) \geq a\, f(x)$, and we get

$$E[X] \geq \sum_{x\geq a} x\, f(x) \geq \sum_{x\geq a} a\, f(x) = a \sum_{x\geq a} f(x)$$

The last part of this is obtained by factoring a out of the sum. So now

$$E[X] \geq a \sum_{x\geq a} f(x)$$

Dividing both sides by the positive number a gives

$$\frac{E[X]}{a} \geq \sum_{x\geq a} f(x)$$

Now all we need to do is recognize that $\sum_{x\geq a} f(x)$ is the same as $\mathrm{P}\{X \geq a\}$ and we have the **Markov inequality**

$$\mathrm{P}\{X \geq a\} \leq \frac{E[X]}{a}$$

for a random variable X that has only nonnegative values. We have shown this for discrete random variables. Techniques from calculus confirm that this inequality is true for continuous random variables.

The Markov inequality works for any nonnegative random variable X. What can we say for a random variable that may have some negative values? We can use instead of X the random variable $(X - \mu)^2$ where μ is the expected value or mean of X. Then $(X - \mu)^2$ is a random variable that is always nonnegative and we can apply the Markov inequality, getting

$$P\left\{(X-\mu)^2 \geq k^2\right\} \leq \frac{E\left[(X-\mu)^2\right]}{k^2}$$

where we use k^2 instead of a and assume that k is positive. The inequality

$$(X-\mu)^2 \geq k^2$$

can be replaced by the inequality

$$|X-\mu| \geq k$$

since both $|X-\mu|$ and k are positive.

Now, remember that $E\left[(X-\mu)^2\right]$ is the variance σ^2 of the random variable X, which was introduced in the previous section, and make a substitution in the preceding inequality, getting the **Chebyshev inequality**

$$P\left\{|X-\mu| \geq k\right\} \leq \frac{\sigma^2}{k^2}$$

which is true for any random variable X with expected value μ and variance σ^2 and any positive number k.

The Markov inequality is useful when only the mean is known and the Chebyshev inequality is useful when only the mean and the variance of a random variable are known. The estimates that they give are not very sharp, but may be the best that can be obtained from so little information about a distribution.

Examples of Application of Inequalities

- Suppose the number of products manufactured each week in a factory is a random variable with mean 100. How likely is it that next week's production will be at least 120?

  First, note that we can use the Markov inequality because the random variable X that gives the number of products is nonnegative and we have the mean μ of X. Substituting 100 for μ and 120 for a in the Markov inequality gives

$$P\{X \geq 120\} \leq \frac{100}{120} \approx 0.833$$

  So $P\{X \geq 120\} \leq 0.833$.

- Suppose the number of products manufactured each week in a factory is a random variable with mean 100 and variance 5. How likely is it that next week's production will be between 90 and 110?

  We can use the Chebyshev inequality because we know the mean and the variance. Substitute 100 for μ and 5 for σ. Use 10 for k since the desired range is within 10 of the mean.

$$P\{|X-100| \geq 10\} \leq \frac{25}{100} = 0.25$$

  Since we want to know the probability that production will be between 90 and 100, we can subtract 0.25 from 1 and get 0.75. So the probability that production will be between 90 and 100 is 0.75.

11.3 The Weak Law of Large Numbers

Suppose we flip a coin 10 times and count the number of heads. Then we do this over and over. For each experiment consisting of 10 flips, the expected value and the variance are the same. We know that for each trial of 10 flips, we could get very different outcomes, anywhere from all heads to all tails. But what happens when we do this experiment over and over? It seems that, in the long run, we ought to be getting closer and closer to half heads and half tails.

To understand what happens when we repeat an experiment over and over, we use the laws of large numbers and the central limit theorem. We will study the weak law of large numbers in this section, then study the central limit theorem and the strong law of large numbers in the next two sections.

If we repeat the same experiment of 10 flips over and over, each trial will be independent of the other trials and will have the same expected value μ and the same variance σ^2. To distinguish the trials, we label the random variable by the number of the trial. Thus we have random variables X_1, X_2, X_3, . . ., each with the same expected value μ and the same variance σ^2.

To see what happens in the long run, we look at the average of the random variables:

$$\frac{X_1 + X_2 + \cdots + X_n}{n}$$

The weak law of large numbers will tell us how likely it is that the average value of the outcomes of a number n of different trials will be close to the expected value μ. We use the following expression, where ε is any positive number

$$P\left\{\left|\frac{X_1 + X_2 + \cdots + X_n}{n} - \mu\right| > \varepsilon\right\}$$

This expression gives the probability that the average of the random variables will differ from the mean by more than the number ε. In the long run, for large values of n, it seems like this probability should be small, and this is verified by the weak law of large numbers.

The **weak law of large numbers** for independent random variables X_1, X_2, X_3, . . ., each with the same expected value μ and the same variance σ^2, says that

$$P\left\{\left|\frac{X_1 + X_2 + \cdots + X_n}{n} - \mu\right| > \varepsilon\right\}$$

gets smaller and smaller as n gets larger and larger. In fact, for a positive number ε, no matter how small it is, we can find some very large number n that will make the average of n trials fall within ε of the expected value.

In other words, the weak law of large numbers tells us there is some large n with the property that if we perform n trials of the experiment, the *average* of the n trials is within ε of the expected value.

To see that this is true, we use Chebyshev's inequality from the previous section, which says that for any random variable X with expected value μ and variance σ^2 and any positive number k

$$P\{|X-\mu| \geq k\} \leq \frac{\sigma^2}{k^2}$$

We apply this inequality to a new random variable, the average of the random variables

$$\frac{X_1 + X_2 + \cdots + X_n}{n}$$

In order to apply Chebyshev's inequality to this new random variable, we need to know its expected value and variance.

We saw in section 10.4 (refer to Chapter 10) that the expected value of a sum of independent random variables is the sum of expected values of the random variables, and the expected value of a multiple of a random variable is that multiple of the expected value. Applying these rules, we get

$$\begin{aligned} E\left[\frac{X_1 + X_2 + \cdots + X_n}{n}\right] &= \frac{1}{n} E\left[X_1 + X_2 + \cdots + X\right] \\ &= \frac{1}{n}\left(E[X_1] + E[X_2] + \cdots + E[X_n]\right) \\ &= \frac{1}{n}\left(\mu + \mu + \cdots + \mu\right) \\ &= \frac{1}{n}(n\,\mu) \\ &= \mu \end{aligned}$$

To determine the variance of the average of the random variables, we need two formulas from section 11.1, first, that the variance of the sum of independent random variables is the sum of the variances and, second, that the variance of a multiple of a random variable is the *square* of the multiple of the variance of the random variable. Applying these rules, we get

$$\begin{aligned}
\text{Var}\left[\frac{X_1 + X_2 + \cdots + X_n}{n}\right] &= \frac{1}{n^2}\,\text{Var}\left[X_1 + X_2 + \cdots + X\right] \\
&= \frac{1}{n^2}\left(\text{Var}\left[X_1\right] + \text{Var}\left[X_2\right] + \cdots + \text{Var}\left[X_n\right]\right) \\
&= \frac{1}{n^2}\left(\sigma^2 + \sigma^2 + \cdots + \sigma^2\right) \\
&= \frac{1}{n^2}\left(n\,\sigma^2\right) \\
&= \frac{\sigma^2}{n}
\end{aligned}$$

With the answers to these computations, we can now apply the Chebyshev inequality, getting

$$\text{P}\left\{\left|\frac{X_1 + X_2 + \cdots + X_n}{n} - \mu\right| \geq \varepsilon\right\} \leq \frac{\sigma^2 / n}{\varepsilon^2} = \frac{\sigma^2}{n\varepsilon^2}$$

or

$$\text{P}\left\{\left|\frac{X_1 + X_2 + \cdots + X_n}{n} - \mu\right| \geq \varepsilon\right\} \leq \frac{\sigma^2}{n\varepsilon^2}$$

No matter what value of ε, is chosen, we can make the expression $\sigma^2 / n\varepsilon^2$ as small as we want by choosing a value of n that is very large. Thus we get the weak law of large numbers, which says that for independent random variables X_1, X_2, X_3, . . ., each with the same mean

$$\text{P}\left\{\left|\frac{X_1 + X_2 + \cdots + X_n}{n} - \mu\right| \geq \varepsilon\right\}$$

goes to 0, or has **limit** 0, as n gets larger and larger.

Now let's go back to the experiment of tossing 10 coins. If we do the experiment over and over, we have independent random variables X_1, X_2, X_3, ... In order to apply the weak law of large numbers, we need to know only the expected value.

The expected value is 5, as we can see from the following calculation.

$$\mu = \sum_{i=1}^{10}(1 \times 0.5) + (0 \times 0.5) = 10 \times 0.5 = 5$$

The weak law of large numbers says that

$$P\left\{\left|\frac{X_1 + X_2 + \cdots + X_n}{n} - 5\right| \geq \varepsilon\right\}$$

has limit 0 as n gets larger and larger.

11.4 The Central Limit Theorem

Chebyshev's inequality and the weak law of large numbers give information about the expected value when we do an experiment over and over. What happens to the variance and standard deviation in the long run? The central limit theorem, one of the most important results in probability theory, tells us that the average of independent random variables will look like the normal distribution in the long run.

Suppose, as before, that we perform an experiment over and over, with each trial independent of the others. We have one random variable X_1, X_2, X_3, . . . for each trial. The sum of the random variables $X_1 + X_2 + \cdots + X_n$, is a new random variable, as we have already seen. If the expected value for each trial is μ and the variance is σ^2, then the random variable $X_1 + X_2 + \cdots + X_n$ has expected value $n\mu$ and variance σ^2. The central limit theorem says that this random variable gets closer and closer to a *normal distribution* with expected value $n\mu$ and variance is $n\sigma^2$ as n gets larger and larger.

As we saw in Chapter 9, any normal distribution can be transformed to a standard normal distribution by substituting the random variable x of a normal distribution with expected value μ and variance σ^2 by the random variable $z = (x - \mu) / \sigma$. For the random variable $X_1 + X_2 + \cdots + X_n$ with expected value $n\mu$, variance $n\sigma^2$, and standard deviation $\sigma\sqrt{n}$, this means that

$$\frac{X_1 + X_2 + \cdots + X_n - n\mu}{\sigma\sqrt{n}}$$

gets closer and closer to the standard normal distribution as n gets larger and larger.

We can now state the **central limit theorem**.

Let $X_1, X_2, X_3, \ldots$ be independent random variables, each with the same expected value μ and the same variance σ^2. Then the random variable

$$\frac{X_1 + X_2 + \cdots + X_n - n\mu}{\sigma\sqrt{n}}$$

gets closer and closer to the standard normal distribution as n gets larger and larger.

The central limit theorem gives the approximation

$$P\left\{a \le \frac{X_1 + X_2 + \cdots + X_n - n\mu}{\sigma\sqrt{n}} \le b\right\} \approx P\{a \le \Phi(Z) \le b\}$$

where Z is the standard normal distribution.

Applications of the central limit theorem will be shown in the following examples. In each case, you start with independent random variables $X_1, X_2, X_3, \ldots$, each having the same expected value μ and the same variance σ^2. If you want an approximation for the probability that the sum of the first n of the random variables is in the range $[a, b]$

$$P\{a \le X_1 + X_2 + \cdots + X_n \le b\}$$

The first step is to standardize this probability; in other words, subtract the mean and divide by the standard deviation. This will give a distribution with mean equal to 0 and standard deviation equal to 1. The original probability is the same as the standardized probability

$$P\left\{\frac{a - n\mu}{\sigma} \le \frac{X_1 + X_2 + \cdots + X_n - n\mu}{\sigma} \le \frac{b - n\mu}{\sigma}\right\}$$

By the central limit theorem, you can approximate this probability by the standard normal distribution, getting

$$P\left\{\frac{a - n\mu}{\sigma} \le \frac{X_1 + X_2 + \cdots + X_n - n\mu}{\sigma} \le \frac{b - n\mu}{\sigma}\right\} \approx P\left\{\frac{a - n\mu}{\sigma} \le \Phi(Z) \le \frac{b - n\mu}{\sigma}\right\}$$

Finally, use the techniques of Chapter 9 to compute the probability.

Note that the central limit theorem does not say anything about the distribution of an individual random variable X, but only about the *sum* of many different random variables all having the same expected value and variance.

The central limit theorem is important for two reasons. First, we can use the central limit theorem to find an approximation for a sum of independent random variables as shown in the following examples. Second, the central limit theorem helps us understand why many measurements have a normal distribution. If a measurement is the sum of many different independent random outcomes, then the central limit theorem says that the measurement itself will be close to a normal distribution.

Examples of Applications of the Central Limit Theorem

- Flip a coin 10 times and count the number of heads. What is the probability that the number of heads is more than five? The random variable X gives the number of heads and has expected value 5, using the formula for the expected value of a binomial experiment, which we computed in section 10.2 (see Chapter 10). The variance is $\sigma^2 = 2.5$, computed using the formula for the variance of a binomial experiment, which we computed in section 11.1.

  The probability that in 10 trials we get an average of six or more heads is

$$P\left\{\frac{\sum_{i=1}^{10} X_i}{10} \geq 6\right\}$$

  which is the same as

$$P\left\{\sum_{i=1}^{10} X_i \geq 60\right\}$$

  Standardize the sum of the random variables X_i to get

$$P\left\{\sum_{i=1}^{10} X_i \geq 60\right\} = P\left\{\sum_{i=1}^{10} \frac{X_i - 10\times 5}{\sqrt{2.5}\sqrt{10}} \geq \frac{60 - 10\times 5}{\sqrt{2.5}\sqrt{10}}\right\}$$

$$\approx P\left\{\sum_{i=1}^{10} \frac{X_i - 50}{5} \geq \frac{60-50}{5}\right\}$$

$$\approx P\left\{\sum_{i=1}^{10} \frac{X_i - 50}{5} \geq 2\right\}$$

Now we apply the central limit theorem, getting

$$\begin{aligned} P\left\{\sum_{i=1}^{10} \frac{X_i - 50}{5} \geq 2\right\} &\approx P\{Z \geq 2\} \\ &\approx 0.5 - \Phi(2) \\ &\approx 0.5 - 0.4772 \\ &\approx 0.0228 \end{aligned}$$

Thus using the central limit theorem, the probability of getting more than five heads on average after 10 flips is approximately 0.0228.

We could have computed this directly, since this is an example with binomial trials. This gives an answer of about 0.0284, so our estimate is off by about 20%.

- Roll a single die and repeat this experiment many times. Each trial is independent of the others, and the expected value and variance are the same for all trials. We can use the central limit theorem to approximate the probability that the sum of the numbers obtained on 10 throws is at least 30 but no more than 40.

  The expected value for each trial is

  $$E[X] = (1+2+3+4+5+6)/6 = 3.5$$

  and the expected value of 10 trials is 35. The variance for one trial is

  $$\begin{aligned} \mathrm{Var}[X] = \sigma^2 &= E\left[(X-3.5)^2\right] \\ &= \left((1-3.5)^2 + (2-3.5)^2 + (3-3.5)^2 + (4-3.5)^2 + (5-3.5)^2 + (6-3.5)^2\right) / 6 \\ &\approx 2.9167 \end{aligned}$$

  The standard deviation for 10 trials is then $10\sigma^2 \approx 29.167$ and the standard deviation for the sum of 10 trials is $\sqrt{10}\sigma \approx 5.401$. We want to find

  $$P\left\{30 \leq \sum_{i=1}^{10} X_i \leq 40\right\}$$

Standardize this probability, getting

$$P\left\{\frac{30-10\times 3.5}{5.401} \le \frac{\sum_{i=1}^{10} X_i - 10\times 3.5}{5.401} \le \frac{40-10\times 3.5}{5.401}\right\}$$

$$\approx P\left\{-0.926 \le \frac{\sum_{i=1}^{10} X_i - 35}{5.401} \le 0.926\right\}$$

We can approximate this with the central limit theorem.

$$P\left\{-0.926 \le \frac{\sum_{i=1}^{10} X_i - 35}{5.401} \le 0.926\right\} \approx P\{-0.926 \le Z \le 0.926\}$$

$$\approx 2\,\Phi(0.926)$$

$$\approx 0.6455$$

Thus when throwing a die 10 times, the probability of getting a sum that is between 30 and 40 is approximately 0.6455.

- An ecologist has heard that gypsy moths have attacked 20% of the trees in a state park, and she wishes to determine the likelihood that this estimate is correct. She chooses 1,000 trees at random and tests them for gypsy moths. Of the 1,000 trees tested, 162 are infested with gypsy moths. To test the estimate that 20% of the trees have been attacked, we assume that we have made 1,000 trials, each with a probability of 0.20 of "success." The expected value for each trial is 0.20, and for 1,000 trials it is 200. The observed number of trees (162) seems small compared to the expected value, and we ask how likely it is that 162 or fewer trees in a sample will show infestation if 20% of the trees have been infested.

The variance for one trial is

$$\begin{aligned}\mathrm{Var}[X] &= \sigma^2 = E\left[(X-0.2)^2\right] \\ &= (1-0.20)^2 \times 0.20 + (0-0.20)^2 \times 0.80 \\ &= 0.16\end{aligned}$$

For 1,000 trials, the standard deviation is $\sqrt{1000}\,\sigma = \sqrt{1000}\,0.4 \approx 12.649$.

We want to find

$$\mathrm{P}\left\{\sum_{i=1}^{1000} X_i \le 162\right\}$$

Standardize this probability, getting

$$\begin{aligned}\mathrm{P}\left\{\frac{\sum_{i=1}^{1000} X_i - 200}{12.649} \le \frac{162-200}{12.649}\right\} \\ \approx \mathrm{P}\left\{\frac{\sum_{i=1}^{1000} X_i - 200}{12.649} \le -3.004\right\}\end{aligned}$$

Now use the central limit theorem to approximate this.

$$\begin{aligned}\mathrm{P}\left\{\frac{\sum_{i=1}^{1000} X_i - 200}{12.649} \le -3.004\right\} &\approx \mathrm{P}\{Z \le -3.004\} \\ &\approx 0.5 - \Phi(3.004) \\ &\approx 0.5 - 0.4987 \\ &\approx 0.0013\end{aligned}$$

Thus the probability that the ecologist found 162 or fewer infested trees in a sample of 1,000 *and* the rate of infestation is 20% is very unlikely. The ecologist is justified in thinking the report is suspect and should do more investigation.

11.5 The Strong Law of Large Numbers

The strong law of large numbers, like the weak law of large numbers, tells us about the long-term behavior of a series of independent trials with the same expected value and variance.

The strong law of large numbers says that the probability is 1 that the average value of a large number of trials is very close to the expected value. This means that for a sequence of independent random variables $X_1, X_2, X_3, \ldots$ all having the same expected value μ and standard deviation σ, the probability is 1 that in the long run, for very large n, the average

$$\frac{X_1 + X_2 + \cdots + X_n}{n}$$

is very close to the mean μ. In fact, by making n large enough, you can get the average as close to the mean as you like.

11.6 Comparison of Inequalities and Laws

In this chapter, we have studied five different laws or inequalities: the Markov inequality, the Chebyshev inequality, the weak law of large numbers, the central limit theorem, and the strong law of large numbers. Even though we have seen some practical applications of these laws, the main applications of all of these are theoretical—they are used to prove other results in probability and statistics. In this section, we compare and contrast these laws.

The most important general observation is that these laws are all given *as probabilities*. Usually, mathematical laws are given as formulas or equations, but here the results are probabilities. In other words, probability is itself being used as a tool to express the laws of probability.

All of the laws in this chapter give information about a probability distribution based on either one or two pieces of data about the distribution—the expected value or the expected value along with the variance—but not the distribution itself. That is why these results are so valuable; they can be applied in situations where we do not have complete knowledge about a distribution. In particular, they can be applied in statistics, where partial data about a population is collected.

The Markov inequality needs the expected value, while the Chebyshev inequality and the central limit theorem need both the expected value and the variance.

The weak law of large numbers and the strong law of large numbers apply to a sequence of identically distributed random variables with the same expected value and variance. You need to know the expected value, but not the variance as long as it is the same for all of the random variables. This condition is automatic in many of the examples we have studied, such as coin tossing, but must be verified in real-world applications.

The Markov inequality and the Chebyshev inequality give estimates that can be very far from the precise values. Because the estimates are not very good, these inequalities are not used much in applications. The Markov inequality is used mainly to prove the Chebyshev inequality. The Chebyshev inequality has applications in other parts of probability, including the proof of the weak law of large numbers.

The central limit theorem is the most important result in this chapter. It gets its name from the central role that it has in probability theory and statistics. In statistics, where a measurement might correspond to the sum of many different independent random variables, the central limit theorem helps explain the prevalence of the normal distribution and the bell-shaped curve.

The weak law of large numbers and the strong law of large numbers are very much alike in that they tell about the average of a large number of independent random variables. The weak law of large numbers was discovered earlier than the strong law of large numbers and is much easier for mathematicians to prove than the strong law.

The weak law of large numbers and the strong law of large numbers are sometimes called the law of averages. Both tell us that, in the long run, the average of the outcomes of the independent trial will be very close to the expected value. The difference is not in the ultimate outcome, but in what happens on the way there, in how the averages converge to the expected value.

The weak law tells about the probability of a limit, and the strong law tells about the limit of a probability. This difference is subtle and is more obvious with the study of limits and convergence in calculus.

Chapter 11 Summary

The **expected value** of a random variable X with probability distribution $f(x)$ is

$$E[X] = \mu_X = \sum x\, f(x)$$

where the sum is taken over all possible values x of the random variable X.

The **mean** of a set of numbers is the sum of the numbers divided by the number of numbers.

The variance and standard deviation measure the spread of the values of a random variable.

The **variance** of a random variable X is the expected value of the square of the difference between the expected value $E[X] = \mu$ and the values that X can take on. It is denoted Var(X).

$$\text{Var}[X] = E\left[(X-\mu)^2\right]$$

or alternatively

$$\text{Var}\left[X\right] = E\left[X^2\right] - \mu^2$$

The units of the variance are the square of the units of the random variable.

Let X and Y be two independent random variables and let a be any number. Then the variance of a product of the random variable X with a is the product of the variance of X with a^2.

$$\text{Var}(aX) = a^2\, \text{Var}(X)$$

$$\text{Var}(X + Y) = \text{Var}(X) + \text{Var}(Y)$$

The **standard deviation** is the square root of the variance. It has the same units as the random variable.

Mean can be compared to the center of gravity of a physical object, variation can be compared to the moment of inertia of an object, and standard distribution can be compared to the radius of gyration of the object.

The **Markov inequality** gives the probability that a random variable X will take on a value larger than some positive number a, which is the probability $P\{X \geq a\}$. The Markov inequality says

$$P\{X \geq a\} \leq \frac{E[X]}{a}$$

The **Chebyshev inequality** estimates the likelihood that a random variable X will take on a value far away from its expected value. For random variable X with expected value μ and variance σ^2 and any positive number k, the Chebyshev inequality says

$$P\left\{\left|X - \mu\right| \geq k\right\} \leq \frac{\sigma^2}{k^2}$$

The **weak law of large numbers** tells how likely it is that the average value of a sequence of independent random variables will be close to the expected value. Let $X_1, X_2, X_3, \ldots$ be independent random variables, each with the same expected value μ and the same variance σ^2. The weak law of large numbers says that the probability, for any positive number ε,

$$P\left\{\left|\frac{X_1 + X_2 + \cdots + X_n}{n} - \mu\right| > \varepsilon\right\}$$

gets smaller and smaller as n gets larger and larger. We say that this probability goes to 0, or has **limit** 0, as n gets larger and larger.

The **central limit theorem**, one of the most important results in probability theory, tells us that the average of a sequence of independent random variables will look like the normal distribution. Let $X_1, X_2, X_3, \ldots$ be independent random variables, each with expected value μ and variance σ^2. Then

$$\frac{X_1 + X_2 + \cdots + X_n - n\mu}{\sigma\sqrt{n}}$$

is a new random variable with expected value 0 and variance 1. The central limit theorem says that this random variable gets closer and closer to the standard normal distribution Z. The central limit theorem gives the approximation

$$P\left\{a \le \frac{X_1 + X_2 + \cdots + X_n - n\mu}{\sigma\sqrt{n}} \le b\right\} \approx P\{a \le \Phi(Z) \le b\}$$

The central limit theorem does not say anything about the distribution of an individual random variable X, but only about the *sum* of many different random variables all having the same expected value and variance.

The strong law of large numbers, like the weak law of large numbers, tells us about the long-term behavior of a series of independent trials with the same expected value and variance.

The **strong law of large numbers** says that the probability is 1 that the average value of a large number of trials is very close to the expected value. Let $X_1, X_2, X_3, \ldots$ be independent random variables, each with the same expected value μ and standard deviation σ. The probability is 1 that in the long run, for very large n, the average

$$\frac{X_1 + X_2 + \cdots + X_n}{n}$$

is very close to the mean μ. In fact, by making n large enough, you can get the average as close to the mean as you like.

Chapter 11 Practice Problems

1. Find the mean of the following sets of numbers.
   - **a.** {2, 5, 8, 44, 50}
   - **b.** {−3, −3, 0, 2, 3, 8}
2. Compute the expected value, variance, and standard deviation for each of the following distributions.

a.

x	10	20	30	40	50
$f(x)$	0.1	0.1	0.6	0.1	0.1

b.

y	5	8	9	12
$g(y)$	0.5	0.3	0.1	0.1

c.

z	20	40	50	60
$h(z)$	0.2	0.2	0.2	0.4

3. The number of eagles observed by an ornithologist each week is a random variable with mean 25. How likely is it that at least 30 eagles will be observed the next week? That at least 40 eagles will be observed?
4. The number of eagles observed by an ornithologist each week is a random variable with mean 25. If we know that the variance is 5, how likely is it that between 20 and 30 eagles will be observed the next week? Between 10 and 40 eagles will be observed?
5. Toss two dice and subtract the smaller number from the larger. Let this difference be the random variable X.
   a. What is the expected value of X?
   b. What is the variance of X?
   c. What is the standard deviation of X?
   d. Repeat this trial 10 times. Use the central limit theorem to estimate the probability that the sum of the outcomes will be at least 20.
6. A quality control inspector has discovered that 10% of the cell phones coming from a production line are defective. The production line is repaired and 200 phones chosen randomly from the next production run are tested. Of these, one is defective. Let X be the random variable giving the number of defectives when one item is tested before the assembly line was repaired.
   a. What is the expected value of X?
   b. What is the variance of X?
   c. What is the standard deviation of X?
   d. Use the central limit theorem to estimate the probability of one defective in 200 trials given a defective rate of 10%.
   e. Do you think the repair of the assembly line was effective?
   f. If 18 of the 200 tested had been defective, how would you evaluate the repair of the assembly line?
7. Suppose you choose a card at random from a standard deck of 52. The random variable X gives the number on the card, counting aces as 1 and face cards as 10. Repeat this experiment so that each draw is from a standard deck of 52 and repeated trials are independent of one another.

For each of the following, identify the result from this chapter that could be used studying the given situation. The five main results from this chapter are the Markov inequality, the Chebyshev inequality, the weak law of large numbers, the central limit theorem, and the strong law of large numbers.

a. What is the distribution of the average value after a large number of trials?

b. What happens to the average value as you do more and more trials?

c. Can you estimate the probability that X will be 10?

d. Can you estimate the probability that the value of X for a given trial will be within 2 of the expected value?

e. Can you estimate the probability that X will be 1 or 2?

8. What do you need to know about the random variable X in order to determine the distribution of repeated trials in the long run?

9. What value does the average value of n trials of an experiment with random variable X approach as n gets very large? How do you know this?

10. What do you need to know about the random variable X in order to estimate the probability that X will be within 3 of its expected value?

11. What do you need to know about the random variable X that only has nonnegative values in order to determine the probability that X will be larger than a given number?

12. How can you determine the probability that the random variable X will be more than 3 away from its expected value after n trials?

Answers to Chapter 11 Practice Problems

1. a. $\frac{2+5+8+44+50}{5} = 21.8$

b. $\frac{-3+-3+0+2+3+8}{6} \approx 1.16667$

2. For these problems, it is easier to use the formula $Var[X] = E[X^2] - \mu^2$.

a. Start by adding the squares of the random variable to the table.

x	10	20	30	40	50
$f(x)$	0.1	0.1	0.6	0.1	0.1
x^2	100	400	900	1600	2500

The expected value μ_X is $(10 \times 0.1) + (20 \times 0.1) + (30 \times 0.6) + (40 \times 0.1) + (50 \times 0.1) = 1 + 2 + 18 + 4 + 5 = 30$.

The variance $Var[X]$ is $E[X^2] - \mu^2 = (100 \times 0.1) + (400 \times 0.1) + (900 \times 0.6) + (1600 \times 0.1) + (2500 \times 0.1) - 900 = 10 + 40 + 540 + 160 + 250 - 900 = 100$.

The standard deviation σ_X is $\sqrt{100} = 10$.

b. Start by adding the squares of the random variable to the table.

y	5	8	9	12
$g(y)$	0.5	0.3	0.1	0.1
y^2	25	64	981	144

The expected value s is $(5 \times 0.5) + (8 \times 0.3) + (9 \times 0.1) + (12 \times 0.1) = 2.5 + 2.4 + 0.9 + 1.2 = 7.0$.

The variance $Var[Y]$ is $E[Y^2] - \mu^2 = (25 \times 0.5) + (64 \times 0.3) + (81 \times 0.1) + (144 \times 0.1) - 49 = 12.5 + 19.2 + 8.1 + 14.4 = 54.2 - 49 = 5.2$.

The standard deviation σ_Y is $\sqrt{5.2} \approx 2.280$.

c. Start by adding the squares of the random variable to the table.

z	20	40	50	60
$h(z)$	0.2	0.2	0.2	0.4
z^2	400	1600	2500	3600

The expected value μ_Z is $(20 \times 0.2) + (40 \times 0.2) + (50 \times 0.2) + (60 \times 0.4) = 4 + 8 + 10 + 24 = 46$.

The variance $Var[Z]$ is $E[Z^2] - \mu^2 = (400 \times 0.2) + (1600 \times 0.2) + (2500 \times 0.2) + (3600 \times 0.4) - 2116 = 80 + 320 + 500 + 1440 - 2116 = 224$.

The standard deviation σ_Z is $\sqrt{224} \approx 14.967$.

3. We can use the Markov inequality because we have the mean of the random variable that gives the number of eagles. Substituting 25 for the mean and 30 or 40 for a in the Markov inequality gives

$$P\{X \geq 30\} \leq \frac{25}{30} \approx 0.833$$

$$P\{X \geq 40\} \leq \frac{25}{40} = 0.625$$

So the probability of observing at least 30 eagles is no more than 0.833 and the probability of observing at least 40 eagles is no more than 0.625.

4. We can use the Chebyshev inequality because the mean is nonnegative, and we know the variance. Use 5 or 15 for k in the formula, since the desired range is within 5 (for 20 to 30 eagles) or within 15 (for 10 to 40 eagles) of the mean.

$$P\{|X-25| \geq 5\} \leq \frac{5}{25} = 0.25$$

$$P\{|X-25| \geq 15\} \leq \frac{5}{225} \approx 0.0222$$

These numbers give the probabilities for observations *outside* the desired range, so we must subtract these values from 1. Thus the probability of observing between 20 and 30 eagles is at least 0.75, and the probability of observing between 10 to 40 eagles is at least 0.9778.

5. **a.** $E[X] = (0 \times 6 + 1 \times 10 + 2 \times 8 + 3 \times 6 + 4 \times 4 + 5 \times 2) \;/\; 36$

$$= 70 \;/\; 36$$

$$\approx 1.944$$

b. $\text{Var}[X] = E\left[X^2\right] - E[X]^2$

$$= \frac{\left(0^2 \times 6 + 1^2 \times 10 + 2^2 \times 8 + 3^2 \times 6 + 4^2 \times 4 + 5^2 \times 2\right)}{36} - \left(\frac{70}{36}\right)^2$$

$$= \frac{210}{36} - \left(\frac{70}{36}\right)^2$$

$$\approx 2.052$$

c. $\sigma \approx \sqrt{2.052} \approx 1.433$

d. The probability that the sum of the outcomes will be at least 22 is

$$P\left\{\sum_{i=1}^{10} X_i \geq 22\right\}$$

Standardize the sum of the random variables X_i to get

$$P\left\{\sum_{i=1}^{10} X_i \geq 22\right\} = P\left\{\sum_{i=1}^{10} \frac{X_i - 10\times 2.052}{1.433\sqrt{10}} \geq \frac{22 - 10\times 2.052}{1.433\sqrt{10}}\right\}$$

$$\approx P\left\{\sum_{i=1}^{10} \frac{X_i - 20.52}{4.532} \geq \frac{1.48}{4.532}\right\}$$

$$\approx P\left\{\sum_{i=1}^{10} \frac{X_i - 20.52}{4.532} \geq 0.3266\right\}$$

Now we can apply the central limit theorem, getting

$$P\left\{\sum_{i=1}^{10} \frac{X_i - 20.52}{4.528} \geq 0.3266\right\} \approx P\{Z \geq 0.3266\}$$

$$\approx 0.5 - \Phi(0.3266)$$

$$\approx 0.5 - 0.1280$$

$$\approx 0.372$$

Thus using the central limit theorem, the probability of getting a sum of outcomes that is at least 20 after 10 throws is approximately 0.37.

6. a. $E[X] = 1\times 0.10 + 0\times 0.90 = 0.10$

b. $\text{Var}[X] = E\left[\left(X-\mu\right)^2\right] = (1-0.10)^2 \times 0.10 + (0-0.10)^2 \times 0.90$

$$= 0.81\times 0.10 + 0.01\times 0.90 = 0.081 + 0.009$$

$$= 0.09$$

or, using the alternative method of finding variance,

$$\text{Var}[X] = E[X^2] - \left(E[X]\right)^2 = \left(1^2\times 0.1 + 0^2\times 0.9\right) - 0.1^2$$

$$= 0.1 - 0.01$$

$$= 0.09$$

c. $\sigma = \sqrt{0.09} = 0.3$

d. We use the central limit theorem applied to a sum of 200 random variables, each with the same expected value 0.10 and variance 0.09 as X. We are interested in the probability

$$P\{X_1 + X_2 + \cdots + X_{200} \le 1\}$$

Standardize this to get

$$P\left\{\frac{X_1 + X_2 + \cdots + X_{200} - 200 \times 0.10}{0.3\sqrt{200}} \le \frac{1 - 200 \times 0.10}{0.3\sqrt{200}}\right\}$$

$$\approx P\left\{\frac{X_1 + X_2 + \cdots + X_{200} - 20}{4.243} \le \frac{-19}{4.243}\right\}$$

$$\approx P\left\{\frac{X_1 + X_2 + \cdots + X_{200} - 20}{4.243} \le -4.478\right\}$$

The central limit theorem says that

$$P\left\{\frac{X_1 + X_2 + \cdots + X_{200} - 20}{4.243} \le -4.478\right\} \approx P\{\Phi(Z) \le -4.478\}$$

$$\approx 0.5 - 0.5$$

$$\approx 0$$

e. Yes. The probability of 10% of the production being defective is close to 0, based on a sample of 200 items.

f. If there had been 18 defective, we would be interested in the probability $P\{X_1 + X_2 + \cdots + X_{200} \le 18\}$.
Standardize this to get

$$P\left\{\frac{X_1 + X_2 + \cdots + X_{200} - 200 \times 0.10}{0.3\sqrt{200}} \le \frac{18 - 200 \times 0.10}{0.3\sqrt{200}}\right\}$$

$$\approx P\left\{\frac{X_1 + X_2 + \cdots + X_{200} - 20}{4.243} \le \frac{-2}{4.243}\right\}$$

$$\approx P\left\{\frac{X_1 + X_2 + \cdots + X_{200} - 20}{4.243} \le -0.471\right\}$$

The central limit theorem says that

$$P\left\{\frac{X_1 + X_2 + \cdots + X_{200} - 20}{4.243} \le -0.471\right\} \approx P\{\Phi(Z) \le -0.471\}$$

$$\approx 0.5 - 0.1817$$

$$\approx 0.3188$$

Now, we are much less confident that the assembly line has been repaired.

7. **a.** The central limit theorem
   b. The strong law of large numbers or the weak law of large numbers
   c. The Markov inequality
   d. The Chebyshev inequality
   e. The Markov inequality
8. You need to know the expected value and the standard deviation so that you can use the central limit theorem.
9. The average value of n trials of an experiment with random variable X approaches the expected value of X as n gets very large. This follows from the strong law of large numbers or the weak law of large numbers.
10. You need to know the expected value and the variance so that you can use the Chebyshev inequality.
11. You need to know the expected value so that you can use the Markov inequality.
12. You can use the weak law of large numbers.

Chapter 12

Markov Chains

The Russian mathematician Andrei Markov, who lived during the late nineteenth and early twentieth centuries, first introduced the idea of a Markov chain to study systems that are subject to random influences. Markov chains model systems that move from one state to another in steps governed by probabilities.

For example, a government agency may want to see how the population in a county changes over time, as people move between country, suburbs, and city parts of the region. This system has three states: country, suburbs, and city. People may seem to move at random, but if we know the proportions of the population who move from one part of the region to another part each year, and these proportions stay the same year by year, a Markov chain can predict how the population changes over time.

For many Markov chains, the long-term behavior is predictable. There are two important cases of long-term behavior. A steady state system gets to a point where each further step leaves the system in approximately the same state. A periodic system cycles through a few different states. If a population were a steady state system, this would mean that the proportion of the population in each of the three regions would eventually remain constant. If it were a periodic system, it would cycle through several different states. This chapter will give us the tools to find the long-term behavior of systems modeled by Markov chains.

12.1 Examples of Markov Chains

We start our study of Markov chains with two representative examples that illustrate the key features of Markov chains, the random walk and a demographic model. In the following sections, we develop the mathematical tools that give insight into Markov chains.

The Random Walk

One of the most famous examples of a Markov chain is the **random walk** or the **drunkard's walk**—a path along a line or in the plane that looks as chaotic as the walk of a drunkard. A random walk on the number line starts at the origin; the first step can be left with probability p and right with probability $(1 - p)$. Each further step has exactly the same likelihood of going to the left or right as the first step. This can be modeled by a Markov chain because the same probabilities determine each step. On the plane, the steps can go left or right, up or down.

Suppose that you start at the origin and flip a coin to determine your next step: heads, you go left one unit, and tails, you go right one unit. After one flip, you could be at 1 or -1; after two flips, you could be at 2, 0, or -2; and so on. After an odd number of steps, you will end up at an odd number, and after an even number of steps, you will end up at an even number.

If we view each step as a Bernoulli trial, then we can analyze a random walk using the techniques from Chapter 7. To land at a number i after n steps, we must go $(n + i) / 2$ steps to the right and $(n - i) / 2$ steps to the left. The probability of this is

$$\binom{n}{\frac{n+i}{2}}(0.5)^{(n+i)/2}(0.5)^{(n-i)/2} = \binom{n}{\frac{n-i}{2}}(0.5)^{n}$$

For example, the probability that you will be at $i = 0$ after $n = 10$ steps is approximately 0.246, and the probability that you will be at $i = 4$ is approximately 0.117. In this example, the probability that you will land on a negative number $-i$ is the same as the probability that you will land on the positive number i by symmetry.

The random walk is important because it is used to model many real world applications, including the Brownian motion of particles in a liquid, the stock market, and the path of a foraging animal.

Demographics

Our next example is from demographics, the study of properties of human populations. We start with a simple example. Suppose that the different regions of Jefferson County can be classified as country, suburbs, or city. Over the past 10 years, data has shown that, on average, for a one-year period, 10% of the country population moves to the suburbs, 5% of the country population moves to the city, 10% of the suburban population moves to the country, 20% of the suburban population moves to the cities, 5% of the city population moves to the country, and 30% of the city population moves to the suburbs. Currently, 50% of the population is in the country, 20% is suburban, and 30% is in the city. We also assume that no one is moving in or out of the county.

To make predictions about the future distribution of the population in the county, we start by interpreting these percentages as probabilities; so if a person lives in the country, the probability that he or she will move to a suburban area is 0.10, the probability of moving to a city area is 0.05, and the probability of staying in the country is $1 - 0.10 - 0.05 = 0.85$. If a person is chosen at random in the county, the probability that he or she lives in the country is 0.50; the probability of living in a suburban area is 0.20; and the probability of living in the city is 0.30.

There are three important ways to organize this information. The first way is in a table, called a **transition table**, shown in Figure 12.1.

		to		
		Country	Suburbs	City
from	Country	0.85	0.10	0.05
	Suburbs	0.10	0.70	0.20
	City	0.05	0.30	0.65

Figure 12.1
Transition table for the population of Jefferson County.

Each entry in the table gives the probability of a person moving from one area or region (corresponding to the row the entry is in) to another region (corresponding to the column the entry is in). The entries in a row corresponding to one of the areas are the probabilities of moving from that area to each of the other two areas or of not moving. This means that the sum of the three entries in each row is 1.00, since every person is accounted for. The entries in a column corresponding to one of the areas are the probabilities of moving *to* that area from each of the other two areas or of staying in that area. These entries usually will not add up to 1.

We can also display this information in a **transition diagram**, shown in Figure 12.2, where circles represent the regions and arrows labeled with numbers represent the probability of an individual moving from one region to another. The arrows from a region to itself indicate the likelihood that an individual stays in that region.

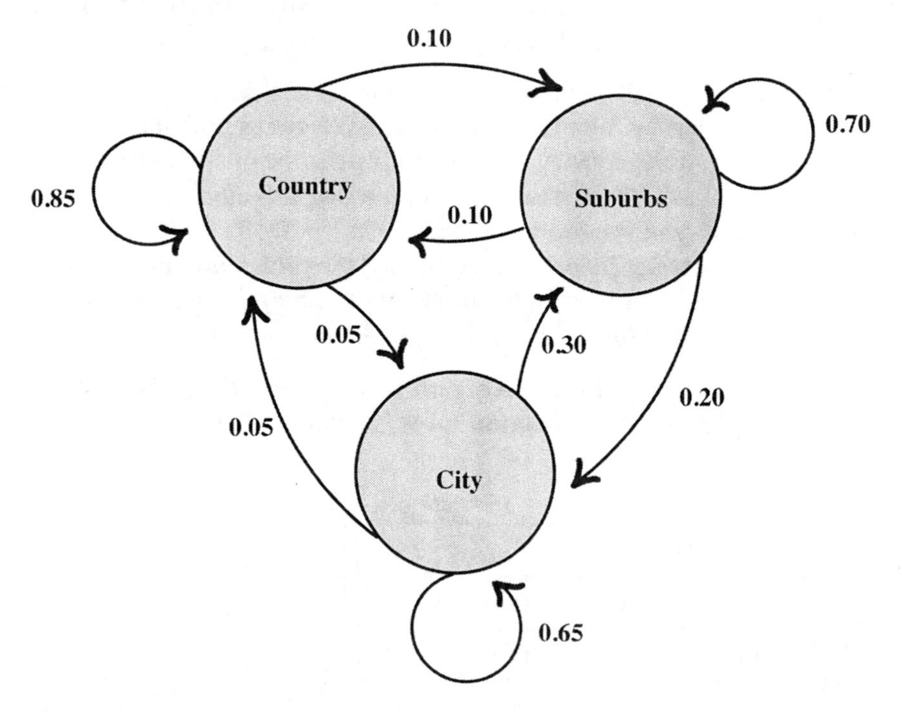

Figure 12.2
Transition diagram.

The transition probabilities can also be given in the form of a **matrix**, a mathematical entity that organizes different numbers into a grid of rows and columns. To construct the transition matrix for this example, we order the regions the same way as in the table shown in Figure 12.1: country, suburbs, city. The entries in a row give the probabilities of a person moving from one area to a new area or staying in the same area. Each column corresponds to the probabilities of a person moving to an area from the other two areas or staying in that area. The transition matrix for this example is

$$\begin{bmatrix} 0.85 & 0.10 & 0.05 \\ 0.10 & 0.70 & 0.20 \\ 0.05 & 0.30 & 0.65 \end{bmatrix}$$

NOTE

Some authors interchange the roles of the rows and columns in the transition matrix. It should be clear from the context if this is the case; then, the sum of the entries in a column is 1. The properties of matrices will be discussed in Section 12.3.

In the following examples, we see how to use this information to answer questions about the population in Jefferson County.

Examples of Demographics Computations

- What percentage of people who live in the city will move in the next year? Since 5% in the city will move to the country and 30% will move to the suburbs, 5% + 30% = 35% of those in the city will move to a different part of the county.
- What percentage of people who live in the suburbs will move? Of those in the suburbs, 10% will move to the country and 20% will move to the city, so 10% + 20% = 30% of those in the suburbs will move to a different part of the county.

- What percentage of people will move to the city? We know that the proportion of the total population that lives in the country and will move to the city is $0.50 \times 0.05 = 0.025$ and the proportion that lives in the suburbs and will move to the city is $0.20 \times 0.20 = 0.04$, and $0.025 + 0.04 = 0.065$, so 6.5% of the population will move to the city.
- What percentage of the whole population will move in the next year? This is more complex because we have to take into account the percentages of the total population in each of the areas. We know 50% of the population lives in the country and $10\% + 5\% = 15\%$ will move, so that is 15% of 50% or 7.5% of the population. Also, 20% now live in the suburbs and 30% will move, $0.20 \times 0.30 = 0.06$, so that gives 6% of the population that will move. Next, 30% of the population is in the city and 35% of those will move, and $0.30 \times 0.35 = 0.105$, so 10.5% will move. So finally, we see that $7.5\% + 6\% + 10.5\% = 24\%$ of the population will move in the next year.

The next two sections review vectors and matrices, and we will see how they can be used to represent demographics models.

Definition of a Markov Chain

The examples of the random walk and the population model illustrate the key features of a Markov chain. First, we have a sequence of random trials. Each of the trials has the same set of possible outcomes, which are called the **states** of the system. In the random walk example, there were two states, taking a step left and taking a step right. In the demographics model, there were three states, country, suburbs, and city. Knowing the probabilities of taking a left or right step or the proportion of the population in each of the regions gives us a **state distribution**.

The distribution in which the system starts is called the **initial state distribution**. For the random walk, the initial state distribution is our starting location, and for the population model, the initial state distribution is the percentage of the population in each region.

The probabilities that tell how likely it is to move from one state to another are called **transition probabilities**. A transition probability gives the likelihood of a move or transition from the current state to any one of the possible states. For the random walk, there are two transition probabilities: a right step or a left step. Each has value equal 0.5.

For the demographics model, there are nine different transition probabilities, since an individual could move from each of the three regions to any one of the three regions.

A **Markov chain** consists of a collection of states along with transition probabilities. Sometimes, we also know the initial state distribution. When you are given a Markov chain and an initial state distribution, your goal is to find the long-term behavior of the system. This long-term behavior is usually given as a collection of probabilities (described in the next section as a state vector), each of which tells the likelihood of being in any one of the states.

In both a Markov chain and a binomial experiment, we perform repeated trials. There is an important distinction, however. In a binomial experiment, as we saw in Chapter 7, the probabilities for the various outcomes are the same in every trial. In a Markov chain, however, the probabilities for the outcome of a trial depend on the result of the previous trial. For example, in the random walk, if our current state is 3, only 2 and 4 are possible outcomes after the next trial.

At best, we can find what the likely distribution of states will be after a long time. This distinction suggests why the study of Markov chains is more difficult and uses the advanced tools of vectors and matrices.

12.2 Vectors

A vector is one object that combines different pieces of data or information. In physics, a vector combines magnitude and direction; an important example is the velocity vector, which combines the speed and direction of travel. A velocity vector is usually shown as an arrow pointing in the direction of travel with length proportional to the speed.

Vectors are very useful in mathematics and its applications because a vector allows you to handle several pieces of information as one mathematical object. This section will present the mathematical properties of vectors that are relevant to the study of Markov chains.

In mathematics, a **vector** is just a list of numbers, usually put inside parentheses. For example, (4, 1), (0, 1, 0), and (1, 3, 0, −1) are all vectors. The numbers themselves are the **components, coordinates,** or **entries** of the vector. Often, we use a lower-case bold letter like ***v*** or ***w*** to represent a vector.

The order of the numbers in the list is important, so the vector (1, 2) is different from the vector (2, 1). Two vectors are the same only when all components are the same and listed in the same order.

Vectors, like numbers, have operations of addition and subtraction. Two vectors can be added or subtracted only when they have the same number of components. Vector addition and subtraction are done componentwise, adding or subtracting corresponding components. For example,

$$(1, 2) + (3, 1) = (1 + 3, 2 + 1) = (4, 3)$$

$$(2, 5, 0, -3) + (0, -1, 8, 3) = (2 + 0, 5 + (-1), 0 + 8, -3 + 3) = (2, 4, 8, 0)$$

$$(1, 2) - (3, 1) = (1 - 3, 2 - 1) = (-2, 1)$$

$$(2, 5, 0, -3) - (0, -1, 8, 3) = (2 - 0, 5 - (-1), 0 - 8, -3 - 3) = (2, 6, -8, -6)$$

A **probability vector** lists the probabilities for all the different outcomes of a random experiment. This means that the entries in a probability vector are all non-negative numbers and that their sum must be equal to 1. Some examples of probability vectors are (1 / 3, 2 / 3), (0.3, 0.25, 0.45), (0.5, 0, 0, 0.5), and (0.1, 0.2, 0.1, 0.3, 0.2, 0.1).

The number of outcomes of the random experiment will be the same as the number of entries in the probability vector.

To give a probability vector, you must know the order that has been given to the outcomes. In the demographics example, the order given was country, suburbs, and city. For an individual living in the country, the probability vector giving the probability of staying in the country or moving to the suburbs or the city is (0.85, 0.10, 0.05). For an individual living in the suburbs, the probability vector is (0.10, 0.70, 0.20) and for an individual living in the city, the probability vector is (0.05, 0.30, 0.65).

A **state distribution vector** gives numbers or probabilities that describe the state a system is in. In the demographics example, the state vector in terms of probabilities is (0.50, 0.20, 0.30). This gives the probability that an individual chosen at random lives in the country, suburbs, or city. In this case, the state vector is also a probability vector.

If we know that the population of the county is 200,000, then the corresponding state vector using numbers will be (100,000, 40,000, 60,000) and gives the initial population for each of the regions. The context will make clear what kind of state vector, probability or numbers, is being used.

To convert a state vector that is given in numbers into a state vector given in probabilities, simply divide each entry in the state vector by the sum of the entries in the state vector. For example, the state vector (3, 3, 2, 2) corresponds to the probability vector (3 / 10, 3 / 10, 2 / 10, 2 / 10) or (0.3, 0.3, 0.2, 0.2).

Suppose that we want to determine how many people there are in the country after one year. There are 100,000 people in the country now, and $0.85 \times 100{,}000 = 85{,}000$ will still be there after one year. There are 40,000 people in the suburbs, and $0.10 \times 40{,}000 = 4{,}000$ of those will have moved to the country after one year. There are 60,000 in the city, and after one year $0.05 \times 60{,}000 = 3{,}000$ of those will have moved to the country. Thus after one year, there will be $85{,}000 + 4{,}000 + 3{,}000 = 92{,}000$ people in the country.

We can summarize this calculation in the formula $(0.85 \times 100{,}000) + (0.10 \times 40{,}000) + (0.05 \times 60{,}000) = 92{,}000$. We can see that the first column of the transition matrix is (0.85, 0.10, 0.05), and the state vector is (100,000, 40,000, 60,000). This sort of computation is done often and is called the **dot product** or the **scalar product** of the two vectors.

The dot product of two vectors $(a, b, c, d) \cdot (x, y, z, w)$ is defined to be the sum of products of corresponding entries, so

$$(a, b, c, d) \cdot (x, y, z, w) = ax + by + cz + dw$$

We can only find the dot product of two vectors if they both have the same number of components. Unlike the sum or difference of two vectors, which is again a vector, the dot product of two vectors is a number, not a vector.

To find out how many people will be in the suburbs after one year using the dot product, we compute

$$(0.10, 0.70, 0.30) \cdot (100{,}000, 40{,}000, 60{,}000)$$
$$= 10{,}000 + 28{,}000 + 18{,}000 = 56{,}000$$

And after one year, again using the dot product, the number of people in the city will be

$$(0.05, 0.20, 0.65) \cdot (100{,}000, 40{,}000, 60{,}000)$$
$$= 5{,}000 + 8{,}000 + 39{,}000 = 52{,}000$$

From these computations, we see that the state vector (100,000, 40,000, 60,000) becomes (92,000 56,000, 52,000) after one year.

12.3 Matrices

Like a vector, a matrix is a single mathematical object that combines different pieces of data or information. The difference is that a matrix is two-dimensional, having rows and columns. The numbers in a row are related to one another in some way, and the numbers in the columns are related to one another in another way.

In the example of a population moving to and from the country, suburbs, and city areas, we had the percentages moving to each of the three areas from a given region in the same row and the percentages moving from each of the three areas to a given region in the same column.

A matrix is given as a grid of numbers enclosed by square brackets. The numbers are arranged in rows and columns, as shown here:

$$\begin{bmatrix} 3 & -1 & 3 \\ 2 & 5 & -3 \\ -4 & 0 & 6 \end{bmatrix}$$

Such a matrix with three rows and three columns is a **three-by-three matrix**. If a matrix has two rows and two columns, it is a **two-by-two matrix**. If a matrix has two rows and three columns, it is a **two-by-three matrix**. Matrices can have any number of rows and columns, and a matrix with n rows and m columns is called an ***n*-by-*m* matrix**. A matrix is **square** if the number of rows is the same as the number of columns.

Because of the way that they are defined, a transition matrix will always be a square matrix, and the number of rows or columns will be the same as the number of states. Matrices are usually represented by upper-case bold letters, like $\boldsymbol{A}$ or $\boldsymbol{B}$.

A vector can be viewed as a matrix with just one row. In such a case, the commas between entries are omitted. So the vector (3, −1, 3) is written as (3 −1 3) if it is considered to be a matrix.

In the previous section, we saw that the dot product of a probability vector of the transitions to a specific region and the current state vector gives the part of the population that will be in the specific region in the following year. This interpretation motivates the rule that tells how to multiply a vector by a matrix.

To multiply a vector by a matrix, form the dot product of the first column with the vector, getting the first component of the answer. Then, form the dot product of the second column with the vector, getting the second component of the answer. Finally form the dot product of the third column with the vector, getting the third component of the answer, as shown here

$$\begin{pmatrix} 100{,}000 & 40{,}000 & 60{,}000 \end{pmatrix} \begin{bmatrix} 0.85 & 0.10 & 0.05 \\ 0.10 & 0.70 & 0.20 \\ 0.05 & 0.30 & 0.65 \end{bmatrix}$$

$$= \big((0.85 \times 100{,}000 + 0.1 \times 40{,}000 + 0.05 \times 60{,}000)$$
$$(0.10 \times 100{,}000 + 0.70 \times 40{,}000 + 0.30 \times 60{,}000)$$
$$(0.05 \times 100{,}000 + 0.20 \times 40{,}000 + 0.65 \times 60{,}000)\big)$$

$$= \big((85{,}000 + 4{,}000 + 3{,}000) \quad (10{,}000 + 28{,}000 + 18{,}000) \quad (5{,}000 + 8{,}000 + 39{,}000) \big)$$

$$= \begin{pmatrix} 92{,}000 & 56{,}000 & 52{,}000 \end{pmatrix}$$

If every row of a matrix is a probability vector, then the matrix could be a transition matrix for some Markov chain and is called a **stochastic matrix**.

Multiplying the state probability vector by the transition matrix gives the state vector at the next stage of the Markov chain. By continuing this process, step by step, using the resulting state as the starting state of the next step, we can determine what the long-term state of the system will be. This is shown next, where we look at Jefferson County after four years.

$$\begin{pmatrix} 100{,}000 & 40{,}000 & 60{,}000 \end{pmatrix} \begin{bmatrix} 0.85 & 0.10 & 0.05 \\ 0.10 & 0.70 & 0.20 \\ 0.05 & 0.30 & 0.65 \end{bmatrix} = \begin{pmatrix} 92{,}000 & 56{,}000 & 52{,}000 \end{pmatrix}$$

$$\begin{pmatrix} 92{,}000 & 56{,}000 & 52{,}000 \end{pmatrix} \begin{bmatrix} 0.85 & 0.10 & 0.05 \\ 0.10 & 0.70 & 0.20 \\ 0.05 & 0.30 & 0.65 \end{bmatrix} = \begin{pmatrix} 86{,}400 & 64{,}000 & 49{,}600 \end{pmatrix}$$

$$\begin{pmatrix} 86{,}400 & 64{,}000 & 49{,}600 \end{pmatrix} \begin{bmatrix} 0.85 & 0.10 & 0.05 \\ 0.10 & 0.70 & 0.20 \\ 0.05 & 0.30 & 0.65 \end{bmatrix} = \begin{pmatrix} 82{,}320 & 68{,}320 & 49{,}360 \end{pmatrix}$$

$$\begin{pmatrix} 82{,}320 & 68{,}320 & 49{,}360 \end{pmatrix} \begin{bmatrix} 0.85 & 0.10 & 0.05 \\ 0.10 & 0.70 & 0.20 \\ 0.05 & 0.30 & 0.65 \end{bmatrix} = \begin{pmatrix} 79{,}272 & 70{,}864 & 48{,}864 \end{pmatrix}$$

Using this model, we see that after four years, the population distribution in Jefferson County will be 79,272 people living in the country, 70,864 living in the suburbs, and 48,864 living in the city.

The operation of multiplying a vector by a matrix can be generalized to the operation of multiplying a matrix by a matrix:

$$\begin{bmatrix} a & b \\ c & d \end{bmatrix} \times \begin{bmatrix} x & y \\ z & w \end{bmatrix} = \begin{bmatrix} ax+bz & ay+bw \\ cx+dz & cy+dw \end{bmatrix}$$

A numerical example with each step shaded is shown in Figure 12.3.

$$\begin{pmatrix} 2 & 1 \\ 3 & -1 \end{pmatrix} \times \begin{pmatrix} -3 & 0 \\ 4 & 1 \end{pmatrix} = \begin{pmatrix} 2\times -3 + 1\times 4 & \\ & \end{pmatrix} = \begin{pmatrix} -2 & \\ & \end{pmatrix}$$

$$\begin{pmatrix} 2 & 1 \\ 3 & -1 \end{pmatrix} \times \begin{pmatrix} -3 & 0 \\ 4 & 1 \end{pmatrix} = \begin{pmatrix} -2 & \\ (-3)\times 3 + 4\times(-1) & \end{pmatrix} = \begin{pmatrix} -2 & \\ -13 & \end{pmatrix}$$

$$\begin{pmatrix} 2 & 1 \\ 3 & -1 \end{pmatrix} \times \begin{pmatrix} -3 & 0 \\ 4 & 1 \end{pmatrix} = \begin{pmatrix} -2 & 0\times 2 + 1\times 1 \\ -13 & \end{pmatrix} = \begin{pmatrix} -2 & 1 \\ -13 & \end{pmatrix}$$

$$\begin{pmatrix} 2 & 1 \\ 3 & -1 \end{pmatrix} \times \begin{pmatrix} -3 & 0 \\ 4 & 1 \end{pmatrix} = \begin{pmatrix} -2 & 1 \\ -13 & 0\times 3 + -1\times 1 \end{pmatrix} = \begin{pmatrix} -2 & 1 \\ -13 & -1 \end{pmatrix}$$

Figure 12.3
Multiplying a matrix by a matrix.

Most graphing calculators and computer algebra software programs can do vector and matrix multiplications.

Properties of Matrix Multiplication

Just as multiplication of numbers has properties that help simplify computations, matrix multiplication has many properties that can be used when working with matrices. These properties include associativity and the existence of an identity element. These properties, along with others that are useful for Markov chains, will be presented in this section.

Associativity of Matrix Multiplication

The associative property of addition or multiplication of numbers says that in a string of operations, it doesn't matter which you do first. So $(2 + 8) + 4 = 2 + (8 + 4) = 14$ and $(3 \times 2) \times 7 = 3 \times (2 \times 7) = 42$. This property holds also for matrix multiplication. If $\boldsymbol{A}$, $\boldsymbol{B}$, and $\boldsymbol{C}$ are matrices, then

$$(\boldsymbol{A} \times \boldsymbol{B}) \times \boldsymbol{C} = \boldsymbol{A} \times (\boldsymbol{B} \times \boldsymbol{C})$$

The associative property of matrix multiplication holds even when, as in the examples we have been using, $\boldsymbol{A}$ is a row matrix and $\boldsymbol{B}$ and $\boldsymbol{C}$ are square matrices.

Repeated Multiplication of a Vector by a Matrix

In looking at the long-term behavior of Markov chains, we will need to make repeated matrix multiplications. Suppose we have a transition matrix $\boldsymbol{A}$ and an initial state vector $\boldsymbol{v}$. Then the following calculations give the first five steps of the evolution of the system.

$$\boldsymbol{vA}$$

$$(\boldsymbol{vA})\boldsymbol{A}$$

$$((\boldsymbol{vA})\boldsymbol{A})\boldsymbol{A}$$

$$(((\boldsymbol{vA})\boldsymbol{A})\boldsymbol{A})\boldsymbol{A}$$

$$((((\boldsymbol{vA})\boldsymbol{A})\boldsymbol{A})\boldsymbol{A})\boldsymbol{A}$$

Since matrix multiplication is associative, these expressions can be simplified to give

$$vA$$
$$vA^2$$
$$vA^3$$
$$vA^4$$
$$vA^5$$

This means that we can multiply the vector v by powers of A to see what the resulting state vector will be. The powers of a matrix are easy to compute using calculators or algebra software.

For example, suppose we have the transition matrix

$$A = \begin{bmatrix} 0.2 & 0.8 \\ 0.7 & 0.3 \end{bmatrix}$$

for a Markov process, and we want to know what the system will be like after five steps for two different initial probability vectors, (0.55, 0.45) or (1, 0). First, we find A^5:

$$A^5 = \begin{bmatrix} 0.45 & 0.55 \\ 0.48125 & 0.51875 \end{bmatrix}$$

Then we can multiply each of the state vectors by this matrix.

$$\begin{pmatrix} 0.55 & 0.45 \end{pmatrix} \begin{bmatrix} 0.45 & 0.55 \\ 0.48125 & 0.51875 \end{bmatrix} = \begin{pmatrix} 0.4640625 & 0.5359375 \end{pmatrix}$$

$$\begin{pmatrix} 1 & 0 \end{pmatrix} \begin{bmatrix} 0.45 & 0.55 \\ 0.48125 & 0.51875 \end{bmatrix} = \begin{pmatrix} 0.45 & 0.55 \end{pmatrix}$$

Identity Matrix

The number 1 has the property that $1 \times a = a \times 1 = a$; we say that 1 is the **identity element** for multiplication of numbers. The matrices

$$\begin{bmatrix} 1 & 0 \\ 0 & 1 \end{bmatrix} \text{ and } \begin{bmatrix} 1 & 0 & 0 \\ 0 & 1 & 0 \\ 0 & 0 & 1 \end{bmatrix}$$

have a similar property for matrices. For two-by-two matrices,

$$\begin{bmatrix} 1 & 0 \\ 0 & 1 \end{bmatrix}\begin{bmatrix} a & b \\ c & d \end{bmatrix} = \begin{bmatrix} a & b \\ c & d \end{bmatrix} = \begin{bmatrix} a & b \\ c & d \end{bmatrix}\begin{bmatrix} 1 & 0 \\ 0 & 1 \end{bmatrix}$$

and for three-by-three matrices,

$$\begin{bmatrix} 1 & 0 & 0 \\ 0 & 1 & 0 \\ 0 & 0 & 1 \end{bmatrix}\begin{bmatrix} a & b & c \\ d & e & f \\ g & h & i \end{bmatrix} = \begin{bmatrix} a & b & c \\ d & e & f \\ g & h & i \end{bmatrix} = \begin{bmatrix} a & b & c \\ d & e & f \\ g & h & i \end{bmatrix}\begin{bmatrix} 1 & 0 & 0 \\ 0 & 1 & 0 \\ 0 & 0 & 1 \end{bmatrix}$$

Any size square matrix with 1s on the diagonal from upper left to lower right and zeros everywhere else is called an **identity matrix**, which is denoted $\boldsymbol{I}$ and has the property that

$$\boldsymbol{MI} = \boldsymbol{M} = \boldsymbol{IM}$$

If the transition matrix of a system is an identity matrix, the system will never change. For example, if there are three states and the initial state is (a, b, c), then

$$\begin{pmatrix} a & b & c \end{pmatrix}\begin{bmatrix} 1 & 0 & 0 \\ 0 & 1 & 0 \\ 0 & 0 & 1 \end{bmatrix} = \begin{pmatrix} a & b & c \end{pmatrix}$$

We see that such a system will always stay in the state (a, b, c).

Properties of Transition Matrices

Transition matrices have several important properties that will help us understand Markov chains and their behavior:

- Transition matrices are square matrices in which the entries of every row add up to 1. This says that every row is a probability vector.
- The product of two transition matrices is a transition matrix. This means that if $\boldsymbol{A}$ and $\boldsymbol{B}$ are transition matrices of the same size, then $\boldsymbol{AB}$ is a transition matrix, also of the same size.
- The power of a transition matrix is a transition matrix. This means that if $\boldsymbol{A}$ is a transition matrix, then $\boldsymbol{A}^n$ is also a transition matrix.

12.4 Long-Term Behavior of Markov Chains

The main goal in the study of Markov chains is to understand the long-term behavior of a system. In this section, we will see how matrix multiplication can help determine what happens in the long term.

First, we need a Markov chain given by a transition matrix. We have seen in earlier sections of this chapter how to translate a verbal description or transition diagram into a transition matrix. We also need the initial state vector.

There are different possibilities for the long-term behavior of a Markov chain. We will study the following two cases:

- The system may eventually reach a specific state, which does not depend on the initial state, and stay there. This state is called a **steady state.**
- The system may cycle through a set of different states. Systems like this are called **periodic**.

Steady States

For example, we can look at the Markov chain with transition matrix

$$D = \begin{bmatrix} 0.3 & 0.7 \\ 0.5 & 0.5 \end{bmatrix}$$

After 10 steps, the transition matrix will be

$$D^{10} = \begin{bmatrix} 0.3 & 0.7 \\ 0.5 & 0.5 \end{bmatrix}^{10} \approx \begin{bmatrix} 0.41667 & 0.58333 \\ 0.41667 & 0.58333 \end{bmatrix}$$

After one more step, the transition matrix will be the same:

$$D^{11} = \begin{bmatrix} 0.3 & 0.7 \\ 0.5 & 0.5 \end{bmatrix}^{11} \approx \begin{bmatrix} 0.41667 & 0.58333 \\ 0.41667 & 0.58333 \end{bmatrix}$$

If the initial state is (0.4, 0.6), then after 10 or 11 steps, the system is in the following state:

$$\begin{pmatrix} 0.4 & 0.6 \end{pmatrix} \begin{bmatrix} 0.41667 & 0.58333 \\ 0.41667 & 0.58333 \end{bmatrix} \approx \begin{pmatrix} 0.41667 & 0.5833 \end{pmatrix}$$

In fact, after 10 or 11 steps, the system is in state (0.41667, 0.5833) *no matter what the initial condition*. This can be seen by the following calculation, where we start with initial probability distribution (p, q). Since this is a probability distribution vector, we know that $p + q = 1$, a fact that is essential for the calculation.

$$\begin{pmatrix} p & q \end{pmatrix}\begin{bmatrix} 0.41667 & 0.58333 \\ 0.41667 & 0.58333 \end{bmatrix} \approx \begin{pmatrix} 0.41667p + 0.41667q & 0.5833p + 0.5833q \end{pmatrix}$$
$$\approx \begin{pmatrix} 0.41667(p+q) & 0.5833(p+q) \end{pmatrix}$$
$$\approx \begin{pmatrix} 0.41667 & 0.5833 \end{pmatrix}$$

In the next calculation, we have a transition matrix with two identical rows and probability distribution vector $(x, 1 - x)$.

$$\begin{pmatrix} x & (1-x) \end{pmatrix}\begin{bmatrix} p & q \\ p & q \end{bmatrix} = \begin{pmatrix} xp + (1-x)p & xq + (1-x)q \end{pmatrix}$$
$$= \begin{pmatrix} (x + (1-x))p & (x + (1-x))q \end{pmatrix} = \begin{pmatrix} p & q \end{pmatrix}$$

This shows that for any transition matrix with rows that are the same and any probability distribution vector, the next step of the Markov chain will be a probability distribution that is the same as a row of the transition matrix.

If a transition matrix $\boldsymbol{M}$ has only positive entries (so there are no zero entries) or if some power of $\boldsymbol{M}$ has only positive entries, then eventually all powers of $\boldsymbol{M}$ will be approximately the same *and* will have all rows equal to one another. Any starting state will eventually get to a state whose probability distribution vector is the same as one row of the transition matrix.

We can go one step further. The next calculation, where $p + q = 1$, shows that if the rows of a transition matrix are all the same, then the probability distribution vector that is equal to a row of the transition matrix will be fixed by the transition matrix. Such a distribution vector is called a **steady state** or **stationary state**.

$$\begin{pmatrix} p & q \end{pmatrix} \begin{bmatrix} p & q \\ p & q \end{bmatrix} = \begin{pmatrix} p^2 + pq & pq + q^2 \end{pmatrix}$$
$$= \begin{pmatrix} p(p+q) & (p+q)q \end{pmatrix}$$
$$= \begin{pmatrix} p & q \end{pmatrix}$$

If a matrix $\boldsymbol{M}$ has a steady state, you can find it by looking at a few high powers of $\boldsymbol{M}$. Usually something like $\boldsymbol{M}^{10}$ will work. If all rows are approximately equal to one another, then you can conclude that the matrix $\boldsymbol{M}$ has a steady state. If not, try a higher power.

Another way to find a steady state for a matrix $\boldsymbol{M}$ is to find a probability vector $\boldsymbol{v}$ such that

$$\boldsymbol{v}\,\boldsymbol{M} = \boldsymbol{v}$$

If we can find such a vector $\boldsymbol{v}$, then also $\boldsymbol{v}\,\boldsymbol{M}^2 = \boldsymbol{v}$, as we can see from this calculation:

$$\boldsymbol{v}\,\boldsymbol{M}^2 = (\boldsymbol{v}\,\boldsymbol{M})\,\boldsymbol{M} = \boldsymbol{v}\,\boldsymbol{M} = \boldsymbol{v}$$

Similarly, any further step $\boldsymbol{v}\,\boldsymbol{M}^n$ will also be equal to $\boldsymbol{v}$. This means that we just need to find a solution $\boldsymbol{v}$ of the equation $\boldsymbol{v}\,\boldsymbol{M} = \boldsymbol{v}$ to get a steady state vector.

For example, using the matrix

$$\boldsymbol{E} = \begin{bmatrix} 0.6 & 0.4 \\ 0.2 & 0.8 \end{bmatrix}$$

with probability vector $\boldsymbol{v}$ given in coordinates by $(x, (1 - x))$ we get the equation

$$\begin{pmatrix} x & (1-x) \end{pmatrix} \begin{bmatrix} 0.6 & 0.4 \\ 0.2 & 0.8 \end{bmatrix} = \begin{pmatrix} 0.6x + 0.2(1-x) & 0.4x + 0.8(1-x) \end{pmatrix}$$
$$= \begin{pmatrix} x & (1-x) \end{pmatrix}$$

To solve this, we write it as a system of equations

$$\begin{cases} 0.6x + 0.2(1-x) = x \\ 0.4x + 0.8(1-x) = 1-x \end{cases}$$

which becomes

$$\begin{cases} 0.4x + 0.2 = x \\ -0.4x + 0.8 = 1-x \end{cases}$$

and finally

$$\begin{cases} -0.6x = -0.2 \\ 0.6x = 0.2 \end{cases}$$

Solving, we get $x = 1/3 \approx 0.333$ and $(1 - x) = 3/3 \approx 0.667$. This is the steady state, as we can see

$$\begin{pmatrix} 0.333 & 0.667 \end{pmatrix} \begin{pmatrix} 0.6 & 0.4 \\ 0.2 & 0.8 \end{pmatrix} \approx \begin{pmatrix} 0.1998+0.1334 & 0.1332+0.5336 \end{pmatrix}$$

$$\approx \begin{pmatrix} 0.333 & 0.667 \end{pmatrix}$$

Summarizing, if a matrix $\boldsymbol{M}$ has a steady state, there are two ways to find it:

Compute powers of $\boldsymbol{M}$ until you find one that has all rows the same. The steady state vector is the same as one of the rows.

Solve $\boldsymbol{v}\,\boldsymbol{M} = \boldsymbol{v}$ for $\boldsymbol{v}$ to find the steady state distribution vector.

If a matrix $\boldsymbol{M}$ has a steady state, any starting distribution will eventually end up at the steady state distribution.

Periodic States

Sometimes in a Markov chain, we get back to the state that we started with. Such a Markov chain is called **periodic**. Let's start by looking at a simple example, the Markov chain with transition matrix

$$\boldsymbol{F} = \begin{bmatrix} 0 & 1 \\ 1 & 0 \end{bmatrix}$$

If we start with probability distribution vector $\boldsymbol{v} = (p, q)$, we get after the first step

$$\boldsymbol{vF} = \begin{pmatrix} p & q \end{pmatrix} \begin{bmatrix} 0 & 1 \\ 1 & 0 \end{bmatrix} = \begin{pmatrix} q & p \end{pmatrix}$$

and after the second step we get

$$(\boldsymbol{vF})\boldsymbol{F} = \begin{pmatrix} q & p \end{pmatrix} \begin{bmatrix} 0 & 1 \\ 1 & 0 \end{bmatrix} = \begin{pmatrix} p & q \end{pmatrix}$$

so we are back where we started. For the initial state vector (p, q), we go next to (q, p) and then back to (p, q).

This example illustrates periodic behavior, where the system cycles through different states in a very regular manner. Here, we have a cycle of two states, (p, q) and (q, p). Note that in this case, $\boldsymbol{F}^2 = \boldsymbol{I}$, where $\boldsymbol{I}$ is the two-by-two identity matrix.

Because we have two states in the cycle, and the second power of the transition matrix $\boldsymbol{F}$ is 2, we say that this Markov chain is periodic with **period 2**.

Sometimes, we will never get a power of $\boldsymbol{M}$ that is the same as the identity, but instead get $\boldsymbol{M}$ back again. For example, the transition matrix

$$\boldsymbol{M} = \begin{bmatrix} 1 & 0 \\ 1 & 0 \end{bmatrix}$$

has square

$$\begin{bmatrix} 1 & 0 \\ 1 & 0 \end{bmatrix}^2 = \begin{bmatrix} 1 & 0 \\ 1 & 0 \end{bmatrix}$$

In this case, we say that $\boldsymbol{M}$ has period 1. In this case, any vector will be a steady state for the matrix $\boldsymbol{M}$.

To determine whether a Markov chain is periodic or not and to find the period, we can choose an initial or starting distribution vector $\boldsymbol{v}$ and compute $\boldsymbol{vM}$, $\boldsymbol{vM}^2$, $\boldsymbol{vM}^3$, . . ., until we get back to the starting vector $\boldsymbol{v}$ or we can compute $\boldsymbol{M}$, $\boldsymbol{M}^2$, $\boldsymbol{M}^3$, . . ., until we get to the identity matrix $\boldsymbol{I}$ or the starting matrix $\boldsymbol{M}$. If $\boldsymbol{vM}^n = \boldsymbol{v}$ or if $\boldsymbol{M}^n = \boldsymbol{I}$, then the Markov chain has period n.

When $M^{n+1} = M$ even if M^n is not equal to the identity matrix I, then the Markov chain is periodic, with repeating cycle

$$vM, vM^2, vM^3, \ldots, vM^n, vM^{n+1} = vM$$

There are n different states in this cycle, so M has period n.

Summarizing, if a Markov chain with transition matrix M has a repeating cycle of n different states, or if the transition matrix has the property that $M^n = I$ (and n is the smallest such number), then we say that M is periodic with period n. If $M^{n+1} = M$, then M is periodic with period n.

Chapter 12 Summary

A **Markov chain** is a system that has a number of different states along with transition probabilities that tell how likely it is to move from one of the states to another. In a Markov chain, the transition probability to the next state depends on the outcome of the previous. This is different from a binomial experiment, where the probability of the next outcome does not depend on the outcome of the previous trial.

The transitions have a set of outcomes, which are called the **states** of the system. The **probability state distribution** or **state distribution** gives the likelihood that the system is in each of the states. The sum of the probabilities in a state distribution is 1. The distribution in which the system starts is called the **initial state distribution**.

A Markov chain can be described using a **transition table**, a **transition diagram**, or a **transition matrix**.

A **random walk** on the number line starts at the origin; the first step can be left with probability p and right with probability $(1 - p)$. The probability of getting to number i after n steps is

$$\binom{n}{\frac{n+i}{2}} (p)^{(n+i)/2} (1-p)^{(n-i)/2}$$

A **vector** is a list of numbers given in parentheses, like (4, 1), (0, 1, 0), and (1, 3, 0, −1). The numbers are the **components, coordinates,** or **entries** of the vector. Two vectors are the same only when all components are the same and listed in the same order.

Vector addition and subtraction are done componentwise, adding or subtracting corresponding components. For example,

$$(1, 2) + (3, 1) = (1 + 3, 2 + 1) = (4, 3)$$

$$\begin{aligned}(2, 5, 0, -3) - (0, -1, 8, 3) &= (2 - 0, 5 - (-1), 0 - 8, -3 - 3) \\ &= (2, 6, -8, -6)\end{aligned}$$

A **probability vector** or **probability distribution vector** lists the probabilities for all the different outcomes of a random experiment. The entries in a probability vector are all non-negative numbers, and their sum is equal to 1.

A **state distribution vector** is a vector that gives numbers or probabilities that describe what state a system is in.

To convert a state vector that is given in numbers into a state vector given in probabilities, divide each entry in the state vector by the sum of the entries in the state vector.

The **dot product** or the **scalar product** of two vectors $(a, b, c, d) \cdot (x, y, z, w)$ is defined to be the sum of the products of corresponding entries,

$$(a, b, c, d) \cdot (x, y, z, w) = ax + by + cz + dw$$

A **matrix** is a grid of numbers enclosed by square brackets. The numbers are arranged in rows and columns:

$$\begin{bmatrix} 3 & -1 & 3 \\ 2 & 5 & -3 \\ -4 & 0 & 6 \end{bmatrix}$$

A matrix with n rows and m columns is an **n-by-m matrix**. A matrix is **square** if the number of rows is the same as the number of columns. Transition matrices are always square and the number of rows or columns is the same as the number of possible states of the system.

To multiply a vector by a matrix, form the dot product of the first column with the vector, getting the first component of the answer. Then form the dot product of the second column with the vector, getting the second component of the answer, and so on.

The operation of multiplying a vector by a matrix can be generalized to the operation of multiplying a matrix by a matrix. Each entry of the product matrix $\boldsymbol{A} \times \boldsymbol{B}$ is the dot product of a column of $\boldsymbol{B}$ with a row of $\boldsymbol{A}$.

If every row of a matrix is a probability vector, the matrix is called a **stochastic matrix**.

Matrix multiplication is associative. This means that if $\boldsymbol{A}$, $\boldsymbol{B}$, and $\boldsymbol{C}$ are matrices, then

$$(\boldsymbol{A} \times \boldsymbol{B}) \times \boldsymbol{C} = \boldsymbol{A} \times (\boldsymbol{B} \times \boldsymbol{C})$$

Vector and matrix operations can be done with calculators or algebra software.

A square matrix with 1s on the diagonal from upper left to lower right and zeros everywhere else is called an **identity matrix** and is denoted $\boldsymbol{I}$. Whenever $\boldsymbol{M}$ has the same size as $\boldsymbol{I}$, $\boldsymbol{MI} = \boldsymbol{IM} = \boldsymbol{M}$.

The product of two transition matrices is a transition matrix of the same size.

The power of a transition matrix is a transition matrix. If $\boldsymbol{A}$ is a transition matrix, then $\boldsymbol{A}^n$ is also a transition matrix.

Different possibilities for the long-term behavior of a Markov chain include

- The system may eventually reach a specific state that does not depend on the initial state and will stay there. This state is called a **steady state**. If a matrix $\boldsymbol{M}$ has a steady state, then some high power, like $\boldsymbol{M}^{10}$, will have all rows approximately equal to one another. The distribution vector of the steady state will be equal to one of the rows of such a high power of $\boldsymbol{M}$. You can also find a steady state for a matrix $\boldsymbol{M}$ by solving the equation $\boldsymbol{v}\,\boldsymbol{M} = \boldsymbol{v}$ for $\boldsymbol{v}$.
- The system may cycle through a set of different states. Systems like this are called **periodic**. If a Markov chain with transition matrix $\boldsymbol{M}$ has a repeating cycle of n different states or if the transition matrix has the property that $\boldsymbol{M}^n = \boldsymbol{I}$ or if $\boldsymbol{v}\boldsymbol{M}^n = \boldsymbol{v}$ (and n is the smallest such number), then $\boldsymbol{M}$ is periodic with period n. If $\boldsymbol{M}^{n+1} = \boldsymbol{M}$ (and n is the smallest such number), then $\boldsymbol{M}$ has period n.

Chapter 12 Practice Problems

1. Suppose you start a random walk at the origin and move left one unit with probability 0.25 and right one unit with probability 0.75.
   a. What is the probability that you will be at 4 after 10 steps?
   b. What is the probability that you will be at 7 after 13 steps?

c. What is the probability that you will be at -4 after 10 steps? (Note that this is not symmetric like the example in section 12.1.)

2. Derek has a restaurant that serves pizza, tacos, and stir-fry. He has noticed that customers change their choices on return visits according to the transition diagram shown in Figure 12.4.

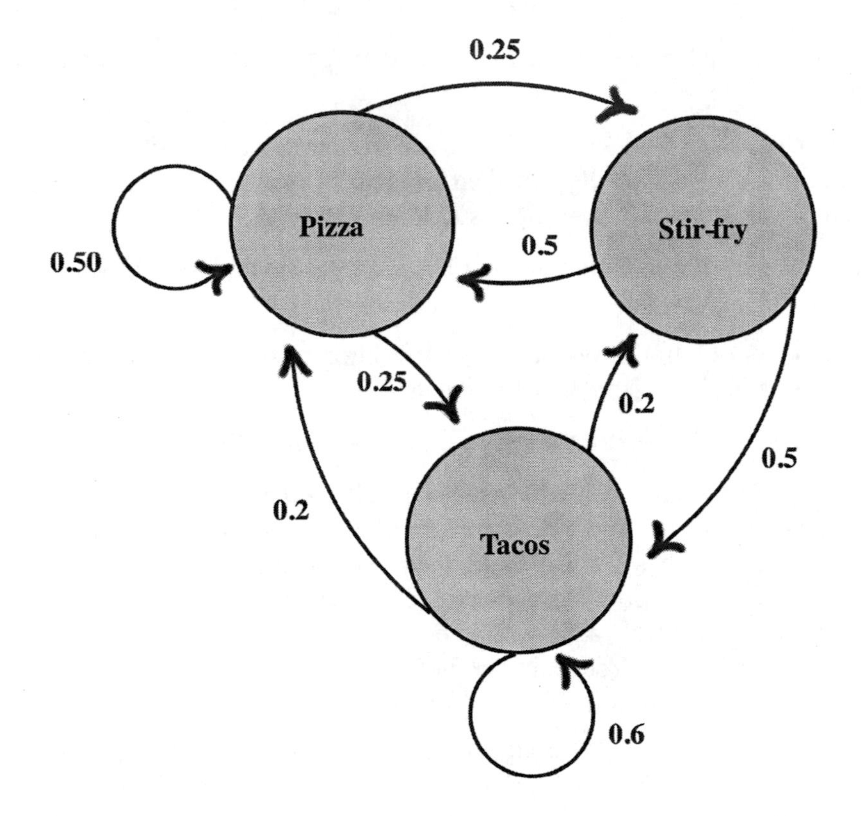

Figure 12.4
Transition diagram for Derek's restaurant.

a. What percentage of people who have pizza order tacos the next time they come?

b. What percentage of people who have tacos order pizza the next time they come?

c. What percentage of people who have pizza order pizza the next time they come?

d. Construct the transition matrix for the transition diagram.

e. Explain what information is contained in the rows and in the columns.

f. On Friday, 140 people order pizza, 50 order tacos, and 70 order stir-fry. What can Derek expect on Saturday?

g. On Tuesday, 50% of the patrons order pizza, 40% order tacos, and 10% order stir-fry. What can he expect on Wednesday?

3. Find the following dot products:

a. $(1, 0) \cdot (3, 8)$

b. $(0.4, 0.3. 0.3) \cdot (20, 50, 300)$

4. Which of the following vectors could be probability vectors? If the vector could not be a probability vector, explain why not.

a. $(0.5, 0.3, 0.2)$

b. $(0.3, 0.4, 0.5)$

c. $(0.2, 0.8)$

d. $(0.2, 0.2, -0.2, 0.2, 0.2)$

e. $(0.2, 0.3, 0.3)$

f. $(0.5, 0.5)$

g. $(0.4, 1.2, 0.4)$

h. $(0.4, 0.2, 0.4)$

i. $(0.2, 1.5, -0.7)$

j. $(0, 0, 0, 1, 0)$

5. Which of the following matrices could be transition matrices? If the matrix could not be a transition matrix, explain why not.

a. $$\begin{bmatrix} 0.2 & 0.5 & 0.3 \\ 0.2 & 0.3 & 0.5 \\ 0.5 & 0.5 & 0 \end{bmatrix}$$

b. $$\begin{bmatrix} 0.3 & 0.3 & 0.3 \\ 0.3 & 0.3 & 0.4 \\ 0.3 & 0.4 & 0.3 \end{bmatrix}$$

c. $$\begin{bmatrix} 0.2 & 0.1 & 0.7 & 0 \\ 0.4 & 0.4 & 0.1 & 0.1 \\ 0.1 & 0.1 & 0.1 & 0.7 \end{bmatrix}$$

d. $\begin{bmatrix} 1 & 0 \\ 0 & .9 \end{bmatrix}$

e. $\begin{bmatrix} 0.8 & 0.8 & -0.6 \\ 0.2 & 0.5 & 0.3 \\ 0 & 1 & 0 \end{bmatrix}$

f. $\begin{bmatrix} 0.2 & 1.8 \\ 0.8 & 0.2 \end{bmatrix}$

6. The transition diagram shown in Figure 12.5 describes a random walk, with steps to the left and steps to the right.

a. Write the transition matrix that corresponds to the transition diagram shown in Figure 12.5.

b. What is the probability that the next step after a step to the right is a step to the left?

c. What is the probability that the next step after a step to the left is a step to the right?

d. What is the probability that the next step after a step to the right is again a step to the right?

e. What is the probability that the next step after a step to the left is again a step to the left?

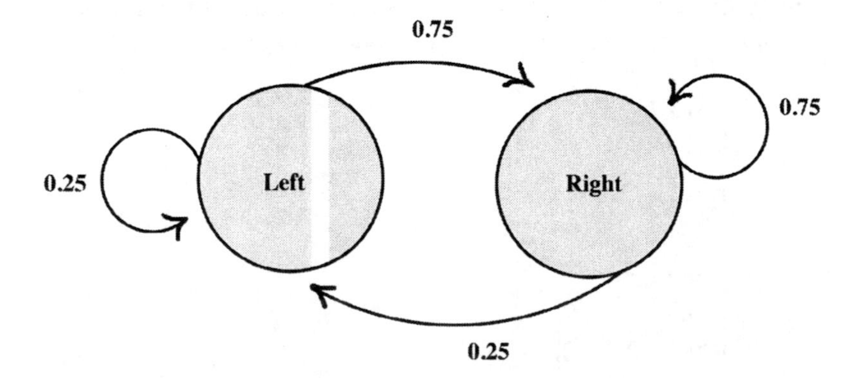

Figure 12.5
Transition diagram for a random walk.

7. Perform the following vector operations:

a. (4, −3) + (2, 2)

b. (−7, 4, 1, 0) + (2, 2, −5, 1)

c. (5, −3) − (0, 2)

d. (4, 1, −1, 5) − (2, 0, 7, 5)

8. Perform the following matrix operations:

a. $\begin{bmatrix} 2 & -2 \\ 1 & 0 \end{bmatrix} + \begin{bmatrix} -1 & 3 \\ 4 & 0 \end{bmatrix}$

b. $\begin{bmatrix} 5 & 5 & 2 \\ 7 & -6 & 1 \\ -5 & 0 & 0 \end{bmatrix} + \begin{bmatrix} -2 & -1 & 3 \\ 0 & 2 & 3 \\ 4 & -4 & 0 \end{bmatrix}$

c. $\begin{bmatrix} 2 & -2 \\ 1 & 0 \end{bmatrix} \times \begin{bmatrix} -1 & 3 \\ 4 & 0 \end{bmatrix}$

d. $\begin{bmatrix} 5 & 5 & 2 \\ 7 & -6 & 1 \\ -5 & 0 & 0 \end{bmatrix} \times \begin{bmatrix} -2 & -1 & 3 \\ 0 & 2 & 3 \\ 4 & -4 & 0 \end{bmatrix}$

9. The following vector operations cannot be performed. For each, give the reason that it cannot be done.

a. (4, 2, 0) + (9, −3)

b. (7, −6) − (6, 5, 0.2)

10. Convert the state vector (4, 12, 3, 1) to a probability vector.

11. The following matrix operations cannot be performed. For each, give the reason that it cannot be done.

a. $\begin{bmatrix} 3 & 5 & 9 & 1 \\ 6 & 1 & 2 & -5 \\ 0 & 1 & -1 & -2 \end{bmatrix} \times \begin{bmatrix} -1 & 0 & 4 \\ 1 & -3 & 7 \\ -4 & 2 & 3 \end{bmatrix}$

b. $\begin{bmatrix} 5 & 3 \\ 4 & -3 \end{bmatrix} \times \begin{bmatrix} 0 & 1 & -1 \\ -4 & 7 & 6 \\ 6 & -3 & 0 \end{bmatrix}$

12. Let matrices A and B be given as follows:

$$A = \begin{bmatrix} 2 & 0 \\ 1 & 1 \end{bmatrix} \text{ and } B = \begin{bmatrix} 3 & 2 \\ -1 & 0 \end{bmatrix}$$

a. Find $A \times B$.

b. Find $B \times A$.

c. What do you notice?

13. Give the five-by-five identity matrix.

14. Perform the following matrix operations. For these operations, you will need a calculator or computer algebra software.

a.
$$\begin{pmatrix} 0.2 & 0.7 & 0.1 \end{pmatrix} \begin{bmatrix} 0.4 & 0.4 & 0.2 \\ 0.6 & 0.1 & 0.3 \\ 0.7 & 0 & 0.3 \end{bmatrix}$$

b.
$$\begin{bmatrix} 0.4 & 0.4 & 0.2 \\ 0.6 & 0.1 & 0.3 \\ 0.7 & 0 & 0.3 \end{bmatrix}^5$$

c.
$$\begin{pmatrix} 0.2 & 0.7 & 0.1 \end{pmatrix} \begin{bmatrix} 0.4 & 0.4 & 0.2 \\ 0.6 & 0.1 & 0.3 \\ 0.7 & 0 & 0.3 \end{bmatrix}^5$$

d.
$$\begin{bmatrix} 0.4 & 0.4 & 0.2 \\ 0.6 & 0.1 & 0.3 \\ 0.7 & 0 & 0.3 \end{bmatrix}^{10}$$

e.
$$\begin{pmatrix} 0.2 & 0.7 & 0.1 \end{pmatrix} \begin{bmatrix} 0.4 & 0.4 & 0.2 \\ 0.6 & 0.1 & 0.3 \\ 0.7 & 0 & 0.3 \end{bmatrix}^{10}$$

f. Make a guess about what this expression will be:

$$\begin{pmatrix} 0.2 & 0.7 & 0.1 \end{pmatrix} \begin{bmatrix} 0.4 & 0.4 & 0.2 \\ 0.6 & 0.1 & 0.3 \\ 0.7 & 0 & 0.3 \end{bmatrix}^{1000}$$

15. The following transition matrices are periodic. Find their periods.

a. $\begin{bmatrix} 0 & 0 & 1 \\ 1 & 0 & 0 \\ 0 & 1 & 0 \end{bmatrix}$

b. $\begin{bmatrix} 0 & 1 & 0 \\ 1 & 0 & 0 \\ 0 & 0 & 1 \end{bmatrix}$

16. Determine whether or not the following transition matrices are periodic. If a matrix is periodic, find its period.

a. $\begin{bmatrix} 1 & 0 \\ 0.5 & 0.5 \end{bmatrix}$

b. $\begin{bmatrix} 0.5 & 0.5 \\ 0.5 & 0.5 \end{bmatrix}$

c. $\begin{bmatrix} 1 & 0 & 0 \\ 0 & 0 & 1 \\ 0 & 1 & 0 \end{bmatrix}$

d. $\begin{bmatrix} 1 & 0 & 0 \\ 0.5 & 0 & 0.5 \\ 0 & 0.5 & 0.5 \end{bmatrix}$

17. Find a stationary state for the transition matrix

$$\begin{bmatrix} 0.4 & 0.6 \\ 0.2 & 0.8 \end{bmatrix}$$

18. Verify that the state vector (0.1, 0.4, 0.5) is a stationary state for the following transition matrix:

$$\begin{bmatrix} 0.2 & 0.5 & 0.3 \\ 0.2 & 0 & 0.8 \\ 0 & 0.7 & 0.3 \end{bmatrix}$$

19. The weather in Sunshine City is either sunny or cloudy. If it is sunny, it is will be sunny the next day with probability 0.8, and it will be cloudy the next day with probability 0.2. If it is cloudy, it will be sunny the next day with probability 0.6 and cloudy the next day with probability 0.4.

a. Draw the transition diagram for weather in Sunshine City.

b. Give the transition matrix for the weather.

c. It is sunny. What is the weather likely to be in three days? In 10 days?

d. It is cloudy. What is the weather likely to be in three days? In 10 days?

e. Is there a stationary state for the weather in Sunshine City?

f. What is the weather like in Sunshine City in the long run?

20. Annika, Bree, and Cheryl like to play cards each week and who wins seems random except no one ever wins two games in a row. However, after Annika wins, it is just as likely that each of her friends win. When Bree wins, Annika wins the next game 80% of the time and Cheryl wins the rest. If Cheryl wins, Annika is likely to win the next game 40% of the time and Bree wins the rest.

a. Draw the transition diagram for the winnings.

b. Give the transition matrix.

c. Annika won the first game this week. After they have played five games, what are the probabilities that each girl will win the next game?

d. Is there a stationary state?

e. What happens in the long run?

21. The study of genetics uses probability to model the transmission of genes from parent to child. Each child inherits two copies of a gene, one from each parent, seemingly at random. In humans, the ability to roll the tongue is governed by one gene, which comes in a dominant (T) or recessive (t) form. An individual's genotype can be TT, Tt, tT, or tt, depending on what is inherited from each parent. Individuals with genotypes TT, Tt, or tT can roll the edges of their tongue. Anyone with genotype tt cannot do this. When two hybrid parents (Tt or tT) have a child, the probability the child is pure dominant (TT) is 0.25, hybrid (Tt or tT) is 0.50, and pure recessive (tt) is 0.25. Suppose a woman marries a hybrid man and they have a child.

a. Fill in the transition matrix, where the rows correspond to the genotype of the mother, and the columns correspond to the genotype of the offspring. The middle row is already filled in because this is the case of a hybrid mother and hybrid father.

$$\begin{array}{c} \\ TT \\ Tt \\ tt \end{array} \begin{array}{c} \begin{array}{ccc} TT & Tt & tt \end{array} \\ \begin{bmatrix} & & \\ 0.25 & 0.50 & 0.25 \\ & & \end{bmatrix} \end{array}$$

b. Suppose a woman marries a hybrid. Their offspring marries a hybrid. What is the probability distribution for a grandchild?

c. What is the probability distribution for a grandchild if the mother has genotype TT?

d. What is the probability distribution for a grandchild if the mother has genotype tt?

e. Is there a steady state if each offspring marries a hybrid?

22. This is a variation of a classic problem in Markov chains. It is very simple to work with, and versions of it can be used to model most Markov chains. There are three urns numbered 1, 2, and 3. In each urn, there are marbles numbered 1, 2, or 3. In the first urn, there are five marbles, of which two are numbered 1, two are numbered 2, and one is numbered 3. In the second urn, there are 10 marbles: three are numbered 1, six are numbered 2, and one is numbered 3. In the third urn, there are eight marbles: four marbles are numbered 1, two are numbered 2, and two are numbered 3. If a marble is drawn from an urn, its number tells which urn to draw from next. It is put back and then the next marble is drawn. This is repeated.

a. Draw the transition diagram for this problem.

b. Give the transition matrix.

c. A marble is drawn from the first urn. What happens next?

d. What happens after two steps?

e. A marble is drawn from the third urn. What happens after 10 steps?

f. Is there a stationary state? What is it?

g. What happens in the long run?

23. Suppose a Markov chain has the following transition matrix:

$$G = \begin{bmatrix} 1 & 0 \\ 0.5 & 0.5 \end{bmatrix}$$

a. Find the stationary vector of G.

b. What is the stationary state of the Markov chain?

c. Is there ever a power of G that has all positive entries?

24. A communications channel transmits bits, which can be either 0 or 1, in several different steps. At each step, there is a chance of error, which depends on the bit being sent. A 0 has a 90% chance of being transmitted correctly and a 1 has a 95% chance of being transmitted correctly.

a. Draw the transition diagram of this situation.

b. Give the transition matrix.

c. Find the stationary distribution vector of this Markov chain.

d. What will happen in the long run, if there are many steps in the communications channel?

Answers to Chapter 12 Practice Problems

1. a. $\binom{10}{7}(0.75)^7(0.25)^3 \approx 0.25028$

b. $\binom{13}{10}(0.75)^{10}(0.25)^3 \approx 0.25165$

c. $\binom{10}{3}(0.75)^3(0.25)^7 \approx 0.0030899$

2. a. 25%

b. 20%

c. 50%

d. Using the order pizza, tacos, and stir-fry, the matrix is

$$\begin{bmatrix} 0.50 & 0.25 & 0.25 \\ 0.20 & 0.60 & 0.20 \\ 0.50 & 0.50 & 0 \end{bmatrix}$$

e. The rows give transition *from* pizza, tacos, and stir-fry in that order and the columns give transition *to* pizza, tacos, and stir-fry in the same order.

f. On Saturday, Derek can expect 115 orders for pizza, 100 orders for tacos, and 45 orders for stir-fry.

g. On Wednesday, Derek can expect 38% will order pizza, 41.5% will order tacos, and 20.5% will order stir-fry.

3. a. $(1, 0) \cdot (3, 8) = 1 \times 3 + 0 \times 8 = 3$

b. $(0.4, 0.3. 0.3) \cdot (20, 50, 300) = 0.4 \times 20 + 0.3 \times 50 + 0.3 \times 300 = 8 + 15 + 90 = 113$

4. a. Yes

b. No, because the sum of the components is not 1.

c. Yes

d. No, because one of the entries is negative and also because the sum of the components is not 1.

e. No, because the sum of the components is not 1.

f. Yes

g. No, because one of the components is greater than 1 and also because the sum of the components is not 1.

h. Yes

i. No, even though the sum of the components is 1, because one of the components is greater than 1 and one of the components is negative.

j. Yes

5. a. Yes

b. No, because the entries in the first row do not add up to 1.

c. No, because the matrix is not square.

d. No, because the entries in the second row do not add up to 1.

e. No, because there is a negative number in the matrix.

f. No, because there is a number greater than 1 in the matrix and the numbers in the first row do not add up to 1.

6. a. Use the top row for transition from the left and the second row for transition from the right. Use the first column for transition to the left and the second column for transition to the right. Then the transition matrix is

$$\begin{bmatrix} 0.25 & 0.75 \\ 0.25 & 0.75 \end{bmatrix}$$

b. 0.25

c. 0.75

d. 0.75

e. 0.25

7. a. $(4, -3) + (2, 2) = (6, -1)$

b. $(-7, 4, 1, 0) + (2, 2, -5, 1) = (-5, 6, -4, 1)$

c. $(5, -3) - (0, 2) = (5, -5)$

d. $(4, 1, -1, 5) - (2, 0, 7, 5) = (2, 1, -8, 0)$

8. a. $$\begin{bmatrix} 2 & -2 \\ 1 & 0 \end{bmatrix} + \begin{bmatrix} -1 & 3 \\ 4 & 0 \end{bmatrix} = \begin{bmatrix} 1 & 1 \\ 5 & 0 \end{bmatrix}$$

b. $$\begin{bmatrix} 5 & 5 & 2 \\ 7 & -6 & 1 \\ -5 & 0 & 0 \end{bmatrix} + \begin{bmatrix} -2 & -1 & 3 \\ 0 & 2 & 3 \\ 4 & -4 & 0 \end{bmatrix} = \begin{bmatrix} 3 & 4 & 5 \\ 7 & -4 & 4 \\ -1 & -4 & 0 \end{bmatrix}$$

c. $$\begin{bmatrix} 2 & -2 \\ 1 & 0 \end{bmatrix} \times \begin{bmatrix} -1 & 3 \\ 4 & 0 \end{bmatrix} = \begin{bmatrix} (2\times-1)+(-2\times4) & (2\times3)+(-2\times0) \\ (1\times-1)+(0\times4) & (1\times3)+(0\times0) \end{bmatrix}$$

$$= \begin{bmatrix} -10 & 6 \\ -1 & 3 \end{bmatrix}$$

d. $$\begin{bmatrix} 5 & 5 & 2 \\ 7 & -6 & 1 \\ -5 & 0 & 0 \end{bmatrix} \times \begin{bmatrix} -2 & -1 & 3 \\ 0 & 2 & 3 \\ 4 & -4 & 0 \end{bmatrix}$$

$$= \begin{bmatrix} -10+0+8 & -5+10-8 & 15+15+0 \\ -14+0+4 & -7-12-4 & 21-18+0 \\ 10+0+0 & 5+0+0 & -15+0+0 \end{bmatrix}$$

$$= \begin{bmatrix} -2 & -3 & 30 \\ -10 & -23 & 3 \\ 10 & 5 & -15 \end{bmatrix}$$

9. a. The two vectors have a different number of components.

b. The two vectors have a different number of components.

10. The state vector (4, 12, 3, 1) corresponds to the probability vector (4 / 20, 12 / 20, 3 / 20, 1 / 20) = (0.20, 0.60, 0.15, 0.05).

11. a. The number of columns in the first matrix is different from the number of rows in the second matrix.

b. The number of columns in the first matrix is different from the number of rows in the second matrix.

12. a. $$A \times B = \begin{bmatrix} 2 & 0 \\ 1 & 1 \end{bmatrix} \begin{bmatrix} 3 & 2 \\ -1 & 0 \end{bmatrix} = \begin{bmatrix} 6 & 4 \\ 2 & 2 \end{bmatrix}$$

b. $$B \times A = \begin{bmatrix} 3 & 2 \\ -1 & 0 \end{bmatrix} \begin{bmatrix} 2 & 0 \\ 1 & 1 \end{bmatrix} = \begin{bmatrix} 8 & 2 \\ -2 & 0 \end{bmatrix}$$

c. $A \times B \neq B \times A$ This is very different from the multiplication of numbers, where $a \times b = b \times a$ for all numbers a and b.

13. $$\begin{bmatrix} 1 & 0 & 0 & 0 & 0 \\ 0 & 1 & 0 & 0 & 0 \\ 0 & 0 & 1 & 0 & 0 \\ 0 & 0 & 0 & 1 & 0 \\ 0 & 0 & 0 & 0 & 1 \end{bmatrix}$$

14. a. $\begin{pmatrix} 0.2 & 0.7 & 0.1 \end{pmatrix} \begin{bmatrix} 0.4 & 0.4 & 0.2 \\ 0.6 & 0.1 & 0.3 \\ 0.7 & 0 & 0.3 \end{bmatrix} = \begin{pmatrix} 0.57 & 0.15 & 0.28 \end{pmatrix}$

b. $\begin{bmatrix} 0.4 & 0.4 & 0.2 \\ 0.6 & 0.1 & 0.3 \\ 0.7 & 0 & 0.3 \end{bmatrix}^5 = \begin{bmatrix} 0.52062 & 0.23148 & 0.24790 \\ 0.52068 & 0.23137 & 0.24795 \\ 0.52073 & 0.23128 & 0.24799 \end{bmatrix}$

c. $\begin{pmatrix} 0.2 & 0.7 & 0.1 \end{pmatrix} \begin{bmatrix} 0.4 & 0.4 & 0.2 \\ 0.6 & 0.1 & 0.3 \\ 0.7 & 0 & 0.3 \end{bmatrix}^5$

$$= \begin{pmatrix} 0.2 & 0.7 & 0.1 \end{pmatrix} \begin{bmatrix} 0.52062 & 0.23148 & 0.24790 \\ 0.52068 & 0.23137 & 0.24795 \\ 0.52073 & 0.23128 & 0.24799 \end{bmatrix}$$

$$= \begin{pmatrix} 0.520673 & 0.231383 & 0.247944 \end{pmatrix}$$

d. $\begin{bmatrix} 0.4 & 0.4 & 0.2 \\ 0.6 & 0.1 & 0.3 \\ 0.7 & 0 & 0.3 \end{bmatrix}^{10} \approx \begin{bmatrix} 0.52066 & 0.23140 & 0.24793 \\ 0.52066 & 0.23140 & 0.24792 \\ 0.52066 & 0.23140 & 0.24793 \end{bmatrix}$

e. $\begin{pmatrix} 0.2 & 0.7 & 0.1 \end{pmatrix} \begin{bmatrix} 0.4 & 0.4 & 0.2 \\ 0.6 & 0.1 & 0.3 \\ 0.7 & 0 & 0.3 \end{bmatrix}^{10}$

$$\approx \begin{pmatrix} 0.2 & 0.7 & 0.1 \end{pmatrix} \begin{bmatrix} 0.52066 & 0.23140 & 0.24793 \\ 0.52066 & 0.23140 & 0.24792 \\ 0.52066 & 0.23140 & 0.24793 \end{bmatrix}$$

$$\approx \begin{pmatrix} 0.52066 & 0.23140 & 0.24793 \end{pmatrix}$$

f. Approximately $\begin{pmatrix} 0.52066 & 0.23140 & 0.24793 \end{pmatrix}$

15. a. $$\begin{bmatrix} 0 & 0 & 1 \\ 1 & 0 & 0 \\ 0 & 1 & 0 \end{bmatrix}^2 = \begin{bmatrix} 0 & 1 & 0 \\ 0 & 0 & 1 \\ 1 & 0 & 0 \end{bmatrix} \text{ and } \begin{bmatrix} 0 & 0 & 1 \\ 1 & 0 & 0 \\ 0 & 1 & 0 \end{bmatrix}^3 = \begin{bmatrix} 1 & 0 & 0 \\ 0 & 1 & 0 \\ 0 & 0 & 1 \end{bmatrix}$$

so the period is 3.

b. $$\begin{bmatrix} 0 & 1 & 0 \\ 1 & 0 & 0 \\ 0 & 0 & 1 \end{bmatrix}^2 = \begin{bmatrix} 1 & 0 & 0 \\ 0 & 1 & 0 \\ 0 & 0 & 1 \end{bmatrix}$$ so the period is 2.

16. a. $$\begin{bmatrix} 1 & 0 \\ 0.5 & 0.5 \end{bmatrix}^2 = \begin{bmatrix} 1 & 0 \\ 0.75 & 0.25 \end{bmatrix} \text{ and}$$

$$\begin{bmatrix} 1 & 0 \\ 0.5 & 0.5 \end{bmatrix}^3 = \begin{bmatrix} 1 & 0 \\ 0.875 & 0.125 \end{bmatrix}$$

This doesn't look periodic, so we look at a much higher power and get

$$\begin{bmatrix} 1 & 0 \\ 0.5 & 0.5 \end{bmatrix}^{10} \approx \begin{bmatrix} 1 & 0 \\ 1 & 0 \end{bmatrix}$$

So this matrix is not periodic. In fact, even though it has at least one 0 in every power, it reaches a stationary state with fixed vector (1, 0).

b. $$\begin{bmatrix} 0.5 & 0.5 \\ 0.5 & 0.5 \end{bmatrix}^2 = \begin{bmatrix} 0.5 & 0.5 \\ 0.5 & 0.5 \end{bmatrix}$$ so this matrix has period 1.

c. $$\begin{bmatrix} 1 & 0 & 0 \\ 0 & 0 & 1 \\ 0 & 1 & 0 \end{bmatrix}^2 = \begin{bmatrix} 1 & 0 & 0 \\ 0 & 1 & 0 \\ 0 & 0 & 1 \end{bmatrix}$$ so this matrix has period 2.

d. $\begin{bmatrix} 1 & 0 & 0 \\ 0.5 & 0 & 0.5 \\ 0 & 0.5 & 0.5 \end{bmatrix}^2 = \begin{bmatrix} 1 & 0 & 0 \\ 0.5 & 0.25 & 0.25 \\ 0.25 & 0.25 & 0.5 \end{bmatrix}$ and

$$\begin{bmatrix} 1 & 0 & 0 \\ 0.5 & 0 & 0.5 \\ 0 & 0.5 & 0.5 \end{bmatrix}^{10} \approx \begin{bmatrix} 1 & 0 & 0 \\ 0.9131 & 0.0332 & 0.0537 \\ 0.8594 & 0.0537 & 0.0869 \end{bmatrix}$$

So after the tenth power, this is starting to look like all rows are the same. Trying a much higher power, we get

$$\begin{bmatrix} 1 & 0 & 0 \\ 0.5 & 0 & 0.5 \\ 0 & 0.5 & 0.5 \end{bmatrix}^{30} \approx \begin{bmatrix} 1 & 0 & 0 \\ 0.999 & 0.000 & 0.001 \\ 0.998 & 0.001 & 0.001 \end{bmatrix}$$

and we can conclude that this matrix is not periodic (but it does have a steady state).

17. Solve the matrix equation $\begin{pmatrix} x & (1-x) \end{pmatrix} = \begin{pmatrix} x & (1-x) \end{pmatrix} \begin{pmatrix} 0.4 & 0.6 \\ 0.2 & 0.8 \end{pmatrix}$

After matrix multiplication, you get a system of two equations:

$$\begin{cases} x = 0.4x + 0.2(1-x) \\ 1-x = 0.6x + 0.8(1-x) \end{cases}$$

The solution is $x = 0.25$ and $(1 - x) = 0.75$. To verify that this solution is correct, perform a matrix multiplication:

$$\begin{pmatrix} 0.25 & 0.75 \end{pmatrix} \begin{bmatrix} 0.4 & 0.6 \\ 0.2 & 0.8 \end{bmatrix} = \begin{pmatrix} 0.1+0.15 & 0.15+0.6 \end{pmatrix} = \begin{pmatrix} 0.25 & 0.75 \end{pmatrix}$$

18. Perform the matrix multiplication:

$$\begin{pmatrix} 0.1 & 0.4 & 0.5 \end{pmatrix} \begin{bmatrix} 0.2 & 0.5 & 0.3 \\ 0.2 & 0 & 0.8 \\ 0 & 0.7 & 0.3 \end{bmatrix}$$

$$= \begin{pmatrix} 0.02+0.08+0 & 0.05+0+0.35 & 0.03+0.32+0.15 \end{pmatrix}$$

$$= \begin{pmatrix} 0.1 & 0.4 & 0.5 \end{pmatrix}$$

This confirms that the given state vector is a stationary vector.

19. a. See Figure 12.6.

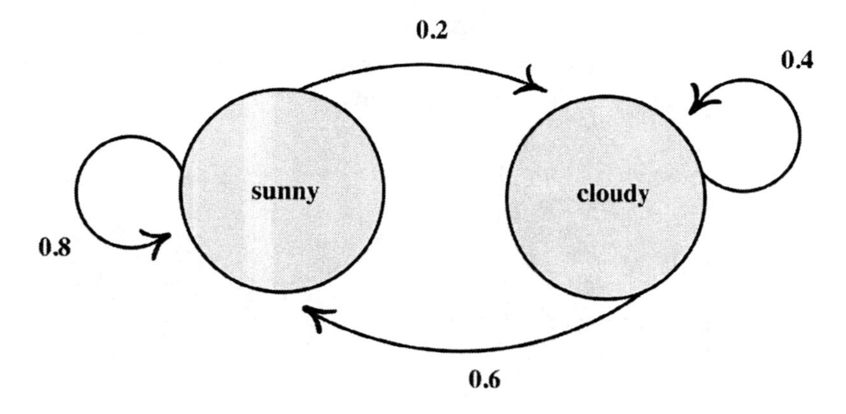

Figure 12.6
Transition diagram for the weather in Sunshine City.

b. $\begin{bmatrix} 0.8 & 0.2 \\ 0.6 & 0.4 \end{bmatrix}$

c. $\begin{pmatrix} 1 & 0 \end{pmatrix} \begin{bmatrix} 0.8 & 0.2 \\ 0.6 & 0.4 \end{bmatrix}^3 = \begin{pmatrix} 0.752 & 0.248 \end{pmatrix}$ Thus three days following a sunny day, there is approximately a 75% chance of sun and a 25% chance of clouds.

$\begin{pmatrix} 1 & 0 \end{pmatrix} \begin{bmatrix} 0.8 & 0.2 \\ 0.6 & 0.4 \end{bmatrix}^{10} \approx \begin{pmatrix} 0.75 & 0.25 \end{pmatrix}$ So 10 days following a sunny day, there is approximately a 75% chance of sun and a 25% chance of clouds.

d. $\begin{pmatrix} 0 & 1 \end{pmatrix} \begin{bmatrix} 0.8 & 0.2 \\ 0.6 & 0.4 \end{bmatrix}^3 = \begin{pmatrix} 0.744 & 0.256 \end{pmatrix}$ So three days following a cloudy day, there is approximately a 75% chance of sun and a 25% chance of clouds.

$\begin{pmatrix} 0 & 1 \end{pmatrix}\begin{bmatrix} 0.8 & 0.2 \\ 0.6 & 0.4 \end{bmatrix}^{10} \approx \begin{pmatrix} 0.75 & 0.25 \end{pmatrix}$ So 10 days following a sunny day, there is approximately a 75% chance of sun and a 25% chance of clouds.

e. Solve $\begin{pmatrix} x & 1-x \end{pmatrix}\begin{bmatrix} 0.8 & 0.2 \\ 0.6 & 0.4 \end{bmatrix} = \begin{pmatrix} x & 1-x \end{pmatrix}$. This is the same as the system

$$\begin{cases} 0.8x + 0.6(1-x) = x \\ 0.2x + 0.4(1-x) = 1-x \end{cases}$$

The solution is $x = 0.75$ and $(1 - x) = 0.25$, so there is a stationary state, (0.75, 0.25),

f. In the long run, there is approximately a 75% chance of sun and a 25% chance of clouds. This explains the name of the city.

20. a. See Figure 12.7.

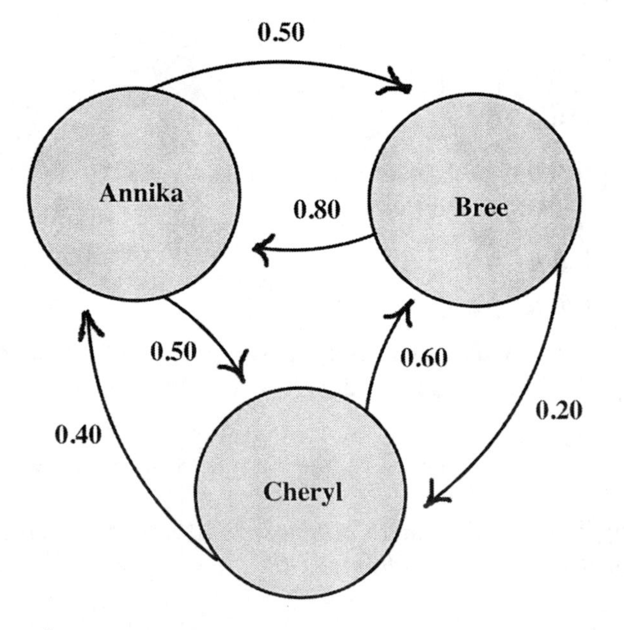

Figure 12.7
Transition diagram for Annika, Bree, and Cheryl.

b. The transition matrix is $\begin{bmatrix} 0 & 0.5 & 0.5 \\ 0.8 & 0 & 0.2 \\ 0.4 & 0.6 & 0 \end{bmatrix}$

c. $$\begin{pmatrix} 1 & 0 & 0 \end{pmatrix} \begin{bmatrix} 0 & 0.5 & 0.5 \\ 0.8 & 0 & 0.2 \\ 0.4 & 0.6 & 0 \end{bmatrix}^5 = \begin{pmatrix} 0.3696 & 0.3432 & 0.2872 \end{pmatrix}$$

After five games, Annika has about a 37% chance of winning the next game, Bree has a 34% chance, and Cheryl has a 29% chance.

d. To find a steady state, solve the equation

$$\begin{pmatrix} x & y & (1-x-y) \end{pmatrix} \begin{bmatrix} 0 & 0.5 & 0.5 \\ 0.8 & 0 & 0.2 \\ 0.4 & 0.6 & 0 \end{bmatrix} = \begin{pmatrix} x & y & (1-x-y) \end{pmatrix}$$

This is the same as the system

$$\begin{cases} 0+0.8y+0.4(1-x-y) = x \\ 0.5x+0y+0.6(1-x-y) = y \\ 0.5x+0.2y+0(1-x-y) = 1-x-y \end{cases}$$

The answer is $x = 22 / 57 \approx 0.386$, $y = 20 / 57 \approx 0.351$, and $(1 - x - y) = 15 /57 \approx 0.263$.

e. In the long run, Annika wins about 38% of the games, Bree wins about 35% of the games, and Cheryl wins about 26% of the games. We can see this from the steady state solution in part d or look at a computation, such as

$$\begin{pmatrix} 1 & 0 & 0 \end{pmatrix} \begin{bmatrix} 0 & 0.5 & 0.5 \\ 0.8 & 0 & 0.2 \\ 0.4 & 0.6 & 0 \end{bmatrix}^{20} \approx \begin{pmatrix} 0.385967 & 0.350876 & 0.263156 \end{pmatrix}$$

21. a.

$$\begin{array}{c|ccc|} & TT & Tt & tt \\ TT & 0.50 & 0.50 & 0 \\ Tt & 0.25 & 0.50 & 0.25 \\ tt & 0 & 0.50 & 0.50 \end{array}$$

b.

$$\begin{pmatrix} 0 & 1 & 0 \end{pmatrix} \begin{bmatrix} 0.50 & 0.50 & 0 \\ 0.25 & 0.50 & 0.25 \\ 0 & 0.50 & 0.50 \end{bmatrix}^2 = \begin{pmatrix} 0 & 1 & 0 \end{pmatrix} \begin{bmatrix} 0.375 & 0.50 & 0.125 \\ 0.25 & 0.50 & 0.25 \\ 0.125 & 0.50 & 0.375 \end{bmatrix}$$
$$= \begin{pmatrix} 0.25 & 0.50 & 0.25 \end{pmatrix}$$

c.

$$\begin{pmatrix} 1 & 0 & 0 \end{pmatrix} \begin{bmatrix} 0.50 & 0.50 & 0 \\ 0.25 & 0.50 & 0.25 \\ 0 & 0.50 & 0.50 \end{bmatrix}^2 = \begin{pmatrix} 1 & 0 & 0 \end{pmatrix} \begin{bmatrix} 0.375 & 0.50 & 0.125 \\ 0.25 & 0.50 & 0.25 \\ 0.125 & 0.50 & 0.375 \end{bmatrix}$$
$$= \begin{pmatrix} 0.375 & 0.50 & 0.125 \end{pmatrix}$$

d.

$$\begin{pmatrix} 0 & 0 & 1 \end{pmatrix} \begin{bmatrix} 0.50 & 0.50 & 0 \\ 0.25 & 0.50 & 0.25 \\ 0 & 0.50 & 0.50 \end{bmatrix}^2 = \begin{pmatrix} 0 & 0 & 1 \end{pmatrix} \begin{bmatrix} 0.375 & 0.50 & 0.125 \\ 0.25 & 0.50 & 0.25 \\ 0.125 & 0.50 & 0.375 \end{bmatrix}$$
$$= \begin{pmatrix} 0.125 & 0.50 & 0.375 \end{pmatrix}$$

e. To determine if there is a steady state over a long period of time, solve the matrix equation

$$\begin{pmatrix} x & y & 1-x-y \end{pmatrix} \begin{bmatrix} 0.50 & 0.50 & 0 \\ 0.25 & 0.50 & 0.25 \\ 0 & 0.50 & 0.50 \end{bmatrix} = \begin{pmatrix} x & y & 1-x-y \end{pmatrix}$$

getting the system of equations

$$\begin{cases} 0.50x + 0.25y = x \\ 0.50x + 0.50y + 0.50(1-x-y) = y \\ 0x + 0.25y + 0.50(1-x-y) = 1-x-y \end{cases}$$

with solution $x = 0.25$, $y = 0.50$, and $(1 - x - y) = 0.25$. We can also look at a high power of the transition matrix. For example, if we look at the tenth power of the transition matrix,

$$\begin{bmatrix} 0.50 & 0.50 & 0 \\ 0.25 & 0.50 & 0.25 \\ 0 & 0.50 & 0.50 \end{bmatrix}^{10} \approx \begin{bmatrix} 0.25 & 0.50 & 0.25 \\ 0.25 & 0.50 & 0.25 \\ 0.25 & 0.50 & 0.25 \end{bmatrix}$$

we get the same probability distribution as when solving the matrix equation. Thus we can conclude that in the long run, under the condition that each offspring marries a hybrid, there will be 25% purebred of genotype TT, 50% hybrid of genotype Tt, and 25% purebred of genotype tt.

22. a. See Figure 12.8.

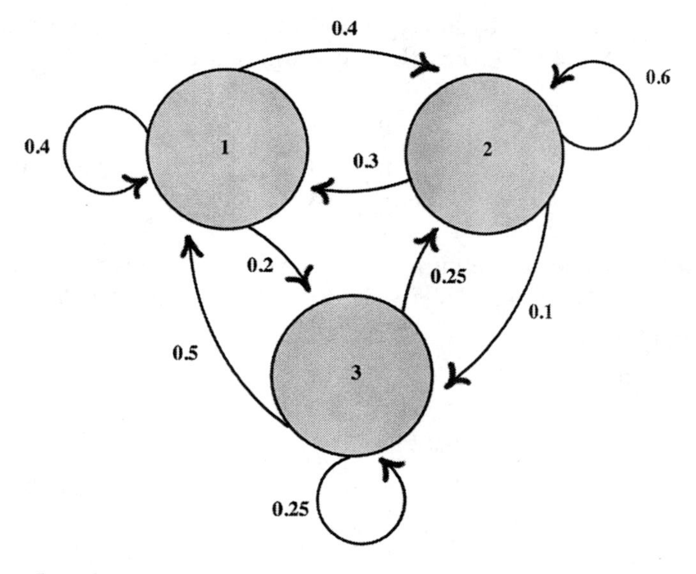

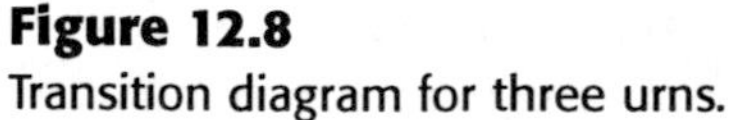

Figure 12.8
Transition diagram for three urns.

b. $$\begin{bmatrix} 0.4 & 0.4 & 0.2 \\ 0.3 & 0.6 & 0.1 \\ 0.5 & 0.25 & 0.25 \end{bmatrix}$$

c. A marble drawn from the first urn has probability distribution vector (1, 0, 0). Multiply this by the transition matrix to get

$$\begin{pmatrix} 1 & 0 & 0 \end{pmatrix} \begin{bmatrix} 0.4 & 0.4 & 0.2 \\ 0.3 & 0.6 & 0.1 \\ 0.5 & 0.25 & 0.25 \end{bmatrix} = \begin{pmatrix} 0.4 & 0.4 & 0.2 \end{pmatrix}$$

This means that for the next draw, the probability of drawing from the first urn is 0.4, from the second it is 0.4, and from the third it is 0.2.

d. Multiply (1, 0, 0) by the square of the transition vector or multiply (0.40, 0.40, 0.20) by the transition matrix to get

$$\begin{pmatrix} 0.4 & 0.4 & 0.2 \end{pmatrix} \begin{bmatrix} 0.4 & 0.4 & 0.2 \\ 0.3 & 0.6 & 0.1 \\ 0.5 & 0.25 & 0.25 \end{bmatrix} = \begin{pmatrix} 0.38 & 0.45 & 0.17 \end{pmatrix}$$

This says that on the third draw, there is a 38% chance of drawing from the first urn, 45% chance from the second, and 17% chance from the third.

e. Compute

$$\begin{pmatrix} 0 & 0 & 1 \end{pmatrix} \begin{bmatrix} 0.4 & 0.4 & 0.2 \\ 0.3 & 0.6 & 0.1 \\ 0.5 & 0.25 & 0.25 \end{bmatrix}^{10} \approx \begin{pmatrix} 0.37 & 0.47 & 0.16 \end{pmatrix}$$

This says that after 10 draws, there will be, for the next draw, approximately a 37% chance of drawing from the first urn, a 47% chance from the second, and a 16% chance from the third.

f. To find a stationary state, solve the equation

$$\begin{pmatrix} x & y & 1-x-y \end{pmatrix} \begin{bmatrix} 0.4 & 0.4 & 0.2 \\ 0.3 & 0.6 & 0.1 \\ 0.5 & 0.25 & 0.25 \end{bmatrix} = \begin{pmatrix} x & y & 1-x-y \end{pmatrix}$$

which corresponds to the system of equation

$$\begin{cases} 0.40x + 0.30y + 0.50(1 - x - y) = x \\ 0.40x + 0.60y + 0.25(1 - x - y) = y \\ 0.20x + 0.10y + 0.25(1 - x - y) = 1 - x - y \end{cases}$$

The solution to this system is $x = 55 / 149$, $y = 70 / 149$, and $(1 - x - y) = 24 / 149$. So the stationary state is approximately (0.3691, 0.4698, 0.1611). So there is a stationary state with this value.

g. In the long run, the probability of drawing from the first urn is about 0.3691, from the second it is about 0.4698, and from the third it is about 0.16107.

23. a. Solve the equation $(x, y)\,\boldsymbol{G} = (x, y)$ to get the stationary vector. The solution is (1, 0).

b. $G^{10} = \begin{bmatrix} 1 & 0 \\ 0.5 & 0.5 \end{bmatrix}^{10} \approx \begin{bmatrix} 1 & 0 \\ 1 & 0 \end{bmatrix}$ and both rows are the same,

so we have the stationary state (1, 0)

c. No, every power of $\boldsymbol{G}$ will have 0 as the upper right-hand entry.

24. a. See Figure 12.9.

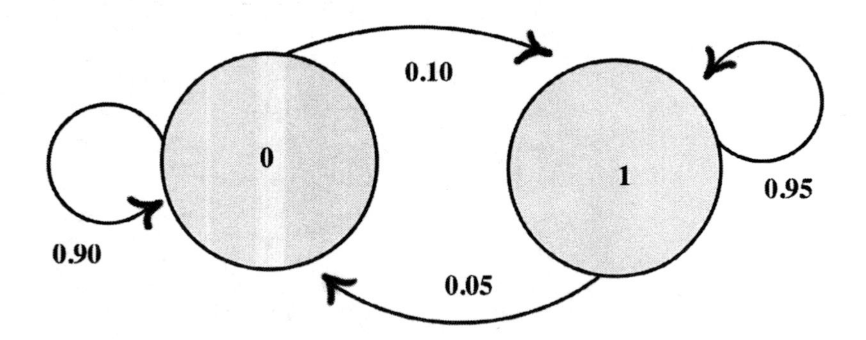

Figure 12.9
Transition diagram for the noisy channel.

b. $\begin{bmatrix} 0.90 & 0.10 \\ 0.05 & 0.95 \end{bmatrix}$

c. Solve the equation $\begin{pmatrix} x & 1-x \end{pmatrix}\begin{bmatrix} 0.90 & 0.10 \\ 0.05 & 0.95 \end{bmatrix} = \begin{pmatrix} x & 1-x \end{pmatrix}$

which is the same as the system

$$\begin{cases} 0.90x + 0.05(1-x) = x \\ 0.10x + 0.95(1-x) = 1-x \end{cases}$$

The solution is $x = 1/3$ and $y = 2/3$.

d. At the end of a communication channel with many steps, the bits will be 1 / 3 0s and 2 / 3 1s, no matter what digits were sent.

Chapter 13

The Axioms of Probability

So far in this book, we have looked at many different aspects of probability and how it can be applied. In this chapter, we will see that probability has its foundation in set theory and mathematical logic. Set theory is based on the very simple idea of the membership relation, as we saw in Chapter 3, and it is the foundation of all of modern mathematics, not just probability. Mathematical logic gives the principles for verifying and validating all of mathematics.

In the 1930s, the Russian mathematician Andrey Kolmogorov developed an axiomatic formulation of probability in terms of set theory. This gave probability theory, even though it deals with randomness and chance, the same rigorous foundation that other areas of mathematics have.

This chapter will describe the axiomatic structure of mathematics, then give the axioms of probability, and then give some examples of theorems derived from the axioms.

13.1 The Axiomatic Structure of Mathematics

Around 300 B.C., the Greek mathematician Euclid lived and taught in the city of Alexandria in Egypt. Actual details of Euclid's life and philosophy have not survived, but in one book, titled simply *The Elements*, Euclid logically and systematically presented all that was then known about geometry. That one book, by laying out a new systematic structure for mathematics, has shaped the development of mathematical theories up until the present day.

The structure that Euclid presented, now known as the **axiomatic structure** of mathematics, began with assumptions. Euclid called them common notions and postulates and treated them as if they were self-evident truths. The common notions, such as equal quantities added to equal quantities give equal quantities, applied to all areas of mathematics. The postulates, such as the construction of a line given two points, were specific to geometry. Today such assumptions or accepted truths are called **axioms.**

Then Euclid gave definitions of important concepts like point, line, plane, circle, and angle. Today mathematicians realize that definitions are limited because they depend on words whose definitions must be known, and so on. Today mathematicians avoid definitions for the most basic terms and instead list them as **undefined terms**, assuming that the reader knows what is being talked about. For example, in geometry, undefined terms include point and line. The axioms give the rules of behavior for undefined terms. Then the theory gives **definitions** for concepts that depend on these more basic undefined terms. In geometry, there are definitions for terms, such as circle, triangle, and square.

With axioms, undefined terms, and definitions, the mathematician sets out to explore the theory. New facts are stated as **propositions** or **theorems**. Then, and this is significant, the theorems are logically derived, using **mathematical proof**, from the axioms and any theorems that have already been proven. This structure assures that all of mathematics has a firm foundation in the assumptions and logic and is reliable.

Thus a mathematical theory consists of the following:

- Undefined terms
- Axioms
- Definitions
- Theorems
- Proofs

Most mathematical theories today rely on set theory, described in Chapter 3, as a foundation. The axioms of probability based on set theory appear in the next section. We base the axioms of probability on set theory so that we can make use of set theory results in probability.

13.2 The Axioms of Probability

The theory of probability starts with a set *S*, called the **sample space**, consisting of elements, which are **simple events**. An **event** *E* is any subset of the sample space *S*.

For each event *E*, there is a number assigned to it, denoted P(*E*). This number is called the **probability of *E***. There are three axioms of probability that describe properties of these probabilities:

Axiom 1 The probability assigned to an event *E* is a number ranging from 0 to 1. In symbols, it would appear as

$$0 \leq \mathrm{P}(E) \leq 1$$

Axiom 2 The sample space includes all outcomes of an experiment or trial. Thus it is a certainty that *something* in the sample space will happen. In terms of probabilities, this says that

$$\mathrm{P}(S) = 1$$

Axiom 3 If events are mutually exclusive, their probabilities add. Recall that events are mutually exclusive if they are disjoint sets (sets with empty intersection). If we have two events, *E* and *F*, this means that

$$\mathrm{P}(E \cup F) = \mathrm{P}(E) + \mathrm{P}(F)$$

This axiom even applies to an infinite sequence of mutually exclusive events, E_1, E_2, E_3, . . .; in this case, the axiom says

$$\mathrm{P}\left(\bigcup_{i=1}^{\infty} E_i\right) = \sum_{i=1}^{\infty} \mathrm{P}\left(E_i\right)$$

It is surprising that only these three simple axioms, together with set theory, give rise to all of probability, but that is one of the beauties of the axiomatic structure of mathematics.

13.3 Theorems Derived from the Axioms of Probability

In this section, we look at three different theorems that follow directly from the axioms of probability theory.

Recall that the complement of a set E is the set of all elements of the universal set S that do not belong to E. The first theorem tells us the probability of the event E^C in terms of the probability of event E. It says that the probability of the complement of an event is obtained by subtracting the probability of the event from 1. For example, the probability of getting a card that is not an ace when we draw one card from a deck of 52 is

$$1 - (4 / 52) = 48 / 52 = 12 / 13$$

Theorem 1 For any event E, $P(E^C) = 1 - P(E)$.

This result follows from the third axiom. The events E and E^C are disjoint, so the third axiom says

$$P(E \cup E^C) = P(E) + P(E^C)$$

but $E \cup E^C = S$ and $P(S) = 1$. By substitution, we get

$$P(E \cup E^C) = 1 = P(E) + P(E^C)$$

and finally

$$P(E^C) = 1 - P(E)$$

The second theorem says that if one event is a subset of another, its probability is no larger than the other. So, for example, the probability of getting a red face card is less than getting a red card.

Theorem 2 If E and F are events such that $E \subset F$, then $P(E) \leq P(F)$.

Here is why this is true. Recall from set theory that if $E \subset F$, then F is made up of two pieces, those elements in E and those elements in F but not in E. See Figure 13.1. In symbols, this is $F = E \cup (E^C F)$.

But by the third axiom, we get the following equation of probabilities:

$$P(F) = P(E \cup (E^C F)) = P(E) + P(E^C F)$$

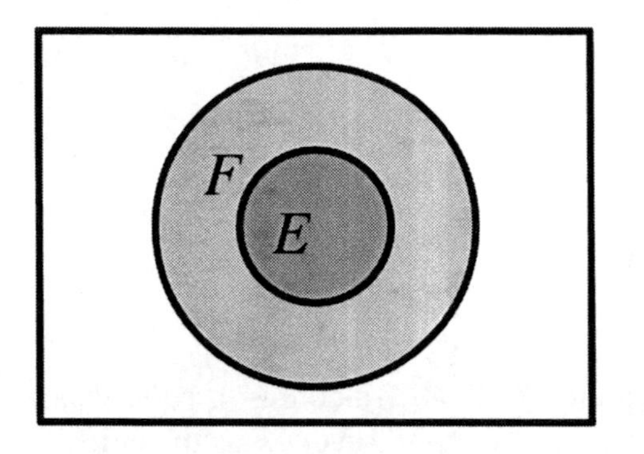

Figure 13.1
Venn diagram for events *E* and *F* with $E \subset F$.

By the first axiom, we know that $0 \leq P(E^C F)$. This means that

$$P(E) \leq P(E) + P(E^C F)$$

and putting this together with the equation $P(F) = P(E) + P(E^C F)$, we get

$$P(E) \leq P(F)$$

The last theorem that we will look at tells what the probability of the union of two events will be. So for two events *E* and *F*, we want to know $P(E \cup F)$. Let's look at the Venn diagram for $E \cup F$, shown in Figure 13.2.

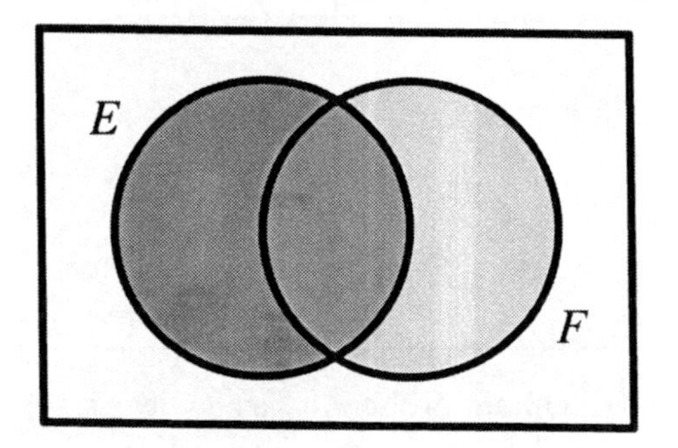

Figure 13.2
Venn diagram for the intersection of events *E* and *F*.

The Venn diagram shows that we can write the event $E \cup F$ as a union of mutually exclusive events

$$E \cup F = E \cup \left(E^C F\right)$$

Use the third axiom to find an expression for $P\left(E \cup F\right)$

$$P\left(E \cup F\right) = P\left(E\right) + P\left(E^C F\right)$$

Now, we want to find an expression for $P\left(E^C F\right)$. To do this, use the Venn diagram in Figure 13.2 to see that we can write the event F as the union of two mutually exclusive events

$$F = EF \cup \left(E^C F\right)$$

and again use the third axiom, getting

$$P\left(F\right) = P\left(EF\right) + P\left(E^C F\right)$$

From this, we see

$$P\left(E^C F\right) = P\left(F\right) - P\left(EF\right)$$

Substitute this expression for $P\left(E^C F\right)$ into the preceding equation that gives $P\left(E \cup F\right)$ to get

$$P\left(E \cup F\right) = P\left(E\right) + P\left(F\right) - P\left(EF\right)$$

This is our third theorem.

Theorem 3 For any two events E and F, $P\left(E \cup F\right) = P(E) + P(F) - P(EF)$.

13.4 Probability Theory

At the beginning of this book, we talked about probability as a way of understanding random behavior. Intuitively, it may seem impossible to use the precise and exact methods of mathematics to study randomness. Using very simple examples of probability, we saw how mathematics can give some good insight into chance events, even though it does not predict what will happen.

Throughout the book, we then developed very powerful mathematical tools for handling probability, including conditional probabilities, random variables, distributions, and expected values. These tools led to much more sophisticated applications.

Finally, in this chapter, we see that the study of random behavior can be put into the axiomatic framework that mathematicians hold to be essential for any mathematical theory. The axiomatic structure of probability provides a firm logical basis for further developments in probability and its applications. Advanced areas of probability include elaborations of the topics we have studied in this book as well as generating functions, stochastic processes, branching processes, uncertainty, and entropy.

Even with the many discoveries made in these advanced topics, the field of probability is still rapidly progressing, providing new tools and insights for applications in all areas of life.

Chapter 13 Summary

The **axiomatic structure** of mathematics consists of

- **Axioms**, which are assumptions or accepted truth.
- **Undefined terms**, which are the most basic terms of the theory.
- **Definitions** for concepts that depend on the more basic undefined terms.
- **Propositions** or **theorems**, which are valid statements about the properties and relationships of the defined and undefined terms.
- **Mathematical proof**, which uses the axioms and theorems that have already been proven.

The axiomatic structure of probability theory is based on set theory and assures that probability has a firm logical foundation and is reliable.

The **sample space**, denoted by S, is a set of elements, which are called **simple events**.

An **event** E is any subset of the sample space S.

For each event E, there is a number assigned to it, denoted $P(E)$, called the **probability of E**.

Axiom 1 The probability assigned to an event E is a number ranging from 0 to 1.

$$0 \leq P(E) \leq 1$$

Axiom 2 The sample space includes all outcomes of an experiment or trial.

$$P(S) = 1$$

Axiom 3 If events are mutually exclusive, their probabilities add. Thus for two events E and F

$$P(E \cup F) = P(E) + P(F)$$

For an infinite sequence of mutually exclusive events, $E_1, E_2, E_3, \ldots$

$$P\left(\bigcup_{i=1}^{\infty} E_i\right) = \sum_{i=1}^{\infty} P(E_i)$$

Theorem 1 For any event E, $P(E^C) = 1 - P(E)$.

Theorem 2 If E and F are two events in the same sample space and $E \subset F$, then $P(E) \leq P(F)$.

Theorem 3 If E and F are two events in the same sample space, then $P(E \cup F) = P(E) + P(F) - P(EF)$.

Chapter 13 Practice Problems

1. Which of the following could be probabilities? If one of the numbers could not be a probability, say why.
   - **a.** 0.6
   - **b.** 1.2
   - **c.** 0.483
   - **d.** −0.36
2. Suppose you are tossing a coin that has been weighted. You discover heads is three times as likely as tails. What is the probability of each event?
3. Let E and F be events in the sample space S with $P(E) = 0.8$ and $P(F) = 0.5$.
   - **a.** What is $P(E^C)$?
   - **b.** What is $P(F^C)$?
   - **c.** What theorem did you use for parts a and b?

d. Could E and F be mutually exclusive events? Explain.

e. Could it be that $E \subset F$?

f. If $P(EF) = 0.6$, what is $P(E \cup F)$?

4. What is $P(\varnothing)$, where $\varnothing$ is the empty set? Justify your answer using the axioms and theorems.

Answers to Chapter 13 Practice Problems

1. a. 0.6 could be a probability

b. 1.2 could not be a probability because it is larger than 1

c. 0.483 could be a probability

d. −0.36 could not be a probability because it is less than 0

2. P(heads) + P(tails) = 1 and P(heads) = 3 P(tails). Therefore P(heads) = 0.75 and P(tails) = 0.25.

3. a. $P(E^C) = 1 - P(E) = 1 - 0.8 = 0.2$

b. $P(F^C) = 1 - P(F) = 1 - 0.5 = 0.5$

c. Theorem 1, which states $P(E^C) = 1 - P(E)$

d. No, E and F could not be mutually exclusive. If they were, the third axiom, which states that, $P(E \cup F) = P(E) + P(F)$ would hold. But then

$$P(E) + P(F) = 0.8 + 0.5 = 1.3$$

would be true. But this cannot be equal to $P(E \cup F)$ since it is larger than 1.

e. No, because if $F \subset E$, then $P(F) \le P(E)$ would also be true. But 0.8 is not less than or
equal to 0.5.

f. $P(E \cup F) = P(E) + P(F) - P(EF) = 0.8 + 0.5 - 0.6 = 0.7$

4. We know that $\varnothing = S^C$. By the first theorem, we can say

$$P(\varnothing) = P(S^C) = 1 - P(S) = 1 - 1 = 0$$

and the answer is $P(\varnothing) = 0$.

Index

D

F

G

H

I

Q

R

S

W

X

Z